# Performance Evaluation and Army Recruiting

T0302970

James N. Dertouzos, Steven Garber

Prepared for the United States Army

ARROYO CENTER

The research described in this report was sponsored by the United States Army under Contract No. W74V8H-06-C-0001.

**Library of Congress Cataloging-in-Publication Data**

Dertouzos, James N., 1950–
     Performance evaluation and Army recruiting / James N. Dertouzos,
Steven Garber.
       p. cm.
     Includes bibliographical references.
     ISBN 978-0-8330-4310-8 (pbk. : alk. paper)
       1. United States. Army—Recruiting, enlistment, etc. 2. United States. Army—
Personnel management.  I. Garber, Steven. II. Title.

     UB323.D455 2006
     355.2'23—dc22

                                                                    2008009719

The RAND Corporation is a nonprofit research organization providing objective analysis and effective solutions that address the challenges facing the public and private sectors around the world. RAND's publications do not necessarily reflect the opinions of its research clients and sponsors.

**RAND®** is a registered trademark.

Published 2008 by the RAND Corporation
1776 Main Street, P.O. Box 2138, Santa Monica, CA 90407-2138
1200 South Hayes Street, Arlington, VA 22202-5050
4570 Fifth Avenue, Suite 600, Pittsburgh, PA 15213-2665
RAND URL: http://www.rand.org
To order RAND documents or to obtain additional information, contact
Distribution Services: Telephone: (310) 451-7002;
Fax: (310) 451-6915; Email: order@rand.org

# Preface

This report documents research methods, findings, and policy conclusions from a project analyzing performance measurement in Army recruiting. The work will interest those involved in the day-to-day management of recruiting resources as well as researchers and analysts engaged in analysis of military enlistment behavior.

This research was sponsored by the Commanding General, U.S. Army Training and Doctrine Command, with U.S. Army Accessions Command as the study lead organization, and was conducted in the Manpower and Training Program of the RAND Arroyo Center. The Arroyo Center is a federally funded research and development center sponsored by the United States Army.

The Project Unique Identification Code (PUIC) for the project that produced this document is ATFCR07222.

For more information on RAND Arroyo Center, contact the Director of Operations (telephone 310-393-0411, extension 6419; FAX 310-451-6952; email Marcy_Agmon@rand.org), or visit Arroyo's Website at http://www.rand.org/ard/.

# Contents

# Figure and Tables

# Summary

Performance metrics are the standard by which individuals and organizations are judged. Such measures are important to organizations because they motivate individuals and influence their choices. In the context of Army recruiting, choices made by recruiters can have a major impact on the ability of the Army to meet its goals. Designing and implementing performance metrics that support Army goals requires analysis of how different metrics would affect recruiter behavior and, in turn, recruiters' contributions toward achieving the Army's goals. In addition, performance measures should not be heavily influenced by random factors affecting enlistment outcomes that might be reasonably attributable to luck or fortune. The present study focuses on performance measurement for Army recruiting to provide incentives that induce behaviors that support achievement of Army goals and are acceptably insensitive to random events.

We compare and evaluate, theoretically and empirically, various performance metrics for regular (i.e., active component) Army recruiting. Some of them have been used by the United States Army Recruiting Command (USAREC); others are original to this study. Previously used performance measures—which we refer to as *traditional* measures—can be computed from readily available data for command or organizational units of various sizes (e.g., stations, companies, battalions) and for intervals of varying lengths. Traditional Army metrics for recruiter performance include the following:

- How many contracts were signed per on-production regular Army (OPRA) recruiter?

- How many high-quality contracts—namely, enlistments of high school seniors and high school graduates scoring in the top half (i.e., categories I to IIIA) of the Armed Forces Qualification Test (AFQT)—were signed per OPRA recruiter?
- By what percentage did the command unit exceed or fall short of recruiting missions (the Army's version of sales targets or quotas) for high-quality or total enlistments?
- How often did the command unit achieve its recruiting targets— the regular Army *mission box*—which, during the period we analyze, included separate targets for (a) high-quality high school graduates, (b) high-quality high school seniors, and (c) other youth?[1]

Our analysis demonstrates that all these measures—and all others that can be computed with readily available data—are flawed because they fail to provide strong incentives (1) for current recruiters to put forth maximum effort or (2) for soldiers who have good skills or aptitudes for recruiting to volunteer for recruiting duty. Moreover, such metrics can be viewed as inequitable; hence, using them can undermine morale and, as a result, reduce the effort levels of recruiters.

Consider an example involving hypothetical recruiting Stations A and B. Suppose that the market territory of Station A is uncommonly fertile for recruiting youth to enlist in the Army. Recruiters in Station A are, in fact, often able to conform to the adage, "make mission, go fishin'." In contrast, the recruiting territory for Station B is uncommonly barren for Army recruiting. Suppose further that USAREC recognizes that Station A has a much better market than does Station B and, in response, assigns to Station A recruiters missions that are double those of Station B. Suppose, finally, that Station A exactly achieves its missions, but that Station B fails to meet its mission. According to all four of the traditional measures described in the bulleted list above, Station A has outperformed Station B. This conclusion, however, is suspect. In particular, to decide which station has really performed

---

[1]  The term *other* refers to the total contracts minus senior high-quality contracts minus graduate high-quality contracts.

better, one must inquire: *How much* better is Station A's market than is Station B's?

Much of the research reported here focused on developing and empirically implementing methods aimed at measuring recruiting performance while taking adequate account of variations in the difficulty of enlisting youth (a) falling into different contract categories (such as high-aptitude seniors versus high school graduates) and (b) located in the market territories of different recruiting stations. Our previous research (Dertouzos and Garber, 2006) demonstrated that the local markets of Army recruiting stations vary significantly in the effort and skill required to enlist high-quality prospects (seniors and graduates combined) and that variations in stations' high-quality missions do not adequately reflect these differences. Thus, performance assessment based on the traditional metrics does not accurately reflect effort exerted and skill applied. In principle, incentives for exerting effort and for persuading the right soldiers to volunteer for recruiting could be stronger.

In Chapter Two, we present a framework for estimating determinants of the numbers of enlistments in various categories that enables estimation of the difficulty of recruiting youth in different categories and in the market areas of different stations. This empirical analysis relies on a microeconomic model (detailed in Appendix A) of recruiter decisions to direct effort toward recruiting youth of different types. Extending previous analyses (Dertouzos and Garber, 2006, Chapter Four), this model emphasizes and distinguishes two general factors that determine recruiting outcomes: recruiter productivity (effort plus skill) and the quality of the organizational unit's market area. We then present a preferred performance metric (PPM) for recruiting stations that explicitly distinguishes among multiple enlistment categories while accounting for variations in local markets that affect the difficulty of enlisting youth who fall into these separate categories. The advantages of this metric come at a cost, however. In particular, implementing them requires econometric analysis to estimate the difficulty of enlisting youth of different types in different local recruiting areas.

In Chapter Three, we present estimates of econometric models of recruiting outcomes using monthly, station-level data for fiscal years 2001 to 2004. The main purposes of our empirical analysis are to quantify the factors that affect the difficulty of recruiting and to use this information to develop estimates of the difficulty of recruiting youth of various types in various locations. We first estimate a model distinguishing the three categories of youth that are missioned separately: high-aptitude, high school graduates; high-aptitude, high school seniors; and "other" enlistments. These estimates provide the empirical foundation for comparing alternative PPMs.

Key findings regarding determinants of enlistments for the three missioned categories of youth include the following:

- At intermediate levels of difficulty of achieving a station's recruiting goal (mission plus Delayed Entry Program (DEP) losses charged that month)—which depend on the level of the goal, the quality of the local market, and other factors—recruiter effort increases as goal difficulty increases.
- The positive marginal effect of increasing goals on effort—and, in turn, on contracts produced—declines as goals increase.
- The marginal effect of goal increases on contract production is substantially higher for stations that have been more successful in recruiting in the recent past. Relying on research literature in psychology and management, we interpret this effect as indicating that success increases recruiters' confidence, their morale, or both.
- Market quality is also an important determinant of recruiter effort levels.
- Market quality in a station's territory depends on many factors, such as qualified military available (QMA) youth per OPRA recruiter, economic and demographic factors, and the strength of competition the Army faces from other services.
- Some determinants of market quality differ substantially across the three missioned categories.
- Actual missions do not adequately reflect differences in market quality. This suggests that enlistment outcomes might be improved through use of performance measures that better reflect such dif-

ferences or by setting missions to more accurately reflect market quality.

We conclude Chapter Three with an exploratory empirical analysis of four enlistment categories that are not missioned separately: (a) high-quality males, (b) other males, (c) high-quality females, and (d) other females. The capability to estimate and account for variations in market quality among these four market segments could be invaluable both for measuring performance and for improving the efficacy of missioning. The analysis uses information on categories of military occupational specialties (MOSs), including combat arms, combat support, blue-collar, and white-collar jobs.[2] Key findings include the following:

- Market quality in the four dimensions varies considerably across station areas.
- Due to variations in local demographics, economic conditions, or both, the difficulty of recruiting youth in one of the four categories provides little, if any, guidance about the difficulty of recruiting youth in other categories. For example, a recruiting station's territories may be difficult for recruiting in one category (e.g., high-quality males) while having an ample supply of prospects in another (e.g., high-quality women).
- Missioning and the distribution of MOSs available to be filled can have important effects on the volume and distribution of enlistments.
- Combat support jobs and white-collar jobs have special appeal to both high-quality men and high-quality women. Combat MOSs are most attractive to lower-quality men; blue-collar jobs draw more "other" women. This finding implies that the distribution

[2] We define MOS categories as follows: (1) *combat arms* MOSs = all occupations that are not available to women; (2) *combat support* MOSs = all other jobs that have no obvious private sector counterpart, such as weapon system maintenance or missile operators; (3) *blue collar* MOSs = jobs with private sector counterparts that are considered blue collar, such as construction, truck maintenance, and transportation jobs; and (4) *white collar* MOSs = jobs with private-sector counterparts in office, service, or professional occupations, such as nurses, clerical, or accounting.

of available occupations can differentially affect the difficulty of recruiting in different local markets.

- Incremental returns associated with adding recruiters diminish. However, additional recruiters can potentially expand the market for lower-quality males and all females. This suggests that the high-quality male market is closer to saturation.

In Chapter Four, we compute three versions of the PPM at the station level for fiscal year (FY) 2004 for 1,417 stations with complete data, and we compare them to five diverse, traditional Army recruiting performance measures. Our key findings:

- The traditional performance measures are not highly correlated with any of the three alternative PPMs. More specifically, the range of rank correlations (across stations) for a full year of station-level performance is 0.42 to 0.68.
- These fairly low correlations indicate that classifying stations on the basis of their ranks on any of the five traditional measures is an unreliable guide for assessing station performance. For example, among the stations that would be ranked among the top 25 percent of performers based on frequency of making regular Army mission box, only 46 percent would be ranked in the top quarter using a PPM.[3] Moreover, about 20 percent of those ranking in the top quarter based on mission-box success would be ranked in the *bottom* half using a PPM.
- In sum, performance evaluation using traditional measures can be very misleading.

In Chapter Five, we consider choice of organizational and temporal units for evaluation. First, we consider alternative *performance windows* or time periods over which performance is assessed (e.g., a month, quarter, or year), using estimates from our model of the three missioned enlistment categories. Key findings include the following:

---

[3]   Alternative versions of the PPM, based on different weighting schemes for combining contract counts in different categories, yielded similar results.

- Randomness or luck is a leading determinant of contract production during a month or a handful of months.
- Aggregating performance over months tends to average out this randomness, and thus makes performance measurement over longer time intervals considerably less subject to randomness.
- For example, during a single month, the proportions of the variation in production levels that are predictable using our estimates are only 0.32, 0.10, and 0.27 for high-aptitude graduates, high-aptitude seniors, and other contracts, respectively.
- Much of this randomness averages out using a performance window of six months, with the proportions of variance explained exceeding 0.65 for both graduates and other enlistments, but only 0.31 for seniors.
- Even over 24 months, considerable randomness remains. Senior contracts are the least predictable, with less than 0.60 of the variance explained by the model.
- Unmeasured factors that are station-specific and persist over time account for 75 percent or more of the unexplained variation that remains—even for two-year performance windows.

Next, we use data for FYs 1999–2001 to evaluate the efficacy of individual-recruiter versus station-level missioning and performance evaluation. The analysis takes advantage of a "natural experiment" made possible due to the sequential conversion of brigades from individual to station missioning. More specifically, two brigades were converted at the beginning of FY 2000, and the other three were converted at the beginning of FY 2001. Key findings include the following:

- Station missioning increased production of high-quality contracts by about 8 percent overall during this time.
- For individual stations, the effect of moving to station missioning depends on the level of mission difficulty. For example, as predicted by a theoretical analysis (detailed in Appendix B), for stations for which missions were unusually easy or unusually difficult, converting to station missioning tended to reduce productivity.

- Converting to station missioning appears to have reduced productivity for about 10 percent of the station-month pairs in our data.

Chapter Six concludes with a discussion of the policy implications of the research findings.

## Implications for Policy

Based on our findings, we believe that the Army should adopt modest, perhaps gradual, changes in the ways that recruiters are evaluated. Although the traditional performance metrics are clearly flawed, we are reluctant to recommend an immediate and wholesale adoption of our preferred performance metric for four major reasons:

1. The current missioning process and associated awards (such as badges, stars, and rings) are a deeply ingrained part of the Army recruiting culture. Sudden, dramatic changes are likely to meet resistance and could undermine recruiter morale and productivity at a time when declines in enlistments would be extremely costly.

2. Relative to traditional measures, the PPM is conceptually complex and requires fairly sophisticated econometric analysis of local markets. Implementation of the PPM would place an additional burden on USAREC and, perhaps more importantly, will not be transparent and intuitive to personnel in the field. Although we believe that this new performance metric would be more equitable, perceptions are sometimes more important than reality.

3. The form of the PPM depends on assumptions about enlistment supply, recruiter behavior, and the relative value of enlistments in different categories. Additional analyses should be conducted before settling on a particular version of the PPM.

4. Our research focused primarily on three markets: high-aptitude seniors, high-aptitude graduates, and all other contracts.

Our exploratory work distinguishing males and females indicates that substantial distinctions exist between these segments that should also be considered in the design of a performance metric. Indeed, there are likely to be other segments—based on education or MOS preferences, for example—that are worth considering.

Despite these caveats, we recommend that USAREC consider some short-term adjustments to its procedures. In particular:

- Improve mission allocation algorithms to reflect variations in market quality and differences in market segments.

Current mission allocations do not do a very good job of adjusting for station-area differences in crucial economic and demographic factors that affect the productivity of recruiter effort and skill.[4] Many of the discrepancies between the PPM and the traditional measures stem from the failure to account adequately for market differences in allocating missions. The pattern of such differences varies by market segment and can also be influenced by command-level policies such as the distribution of MOSs.

- Lengthen the performance window to at least six months or smooth monthly rewards by reducing emphasis on station-level mission accomplishment.

Until at least six months of production have been observed, recruiting outcomes at the station level are dominated by randomness. The importance of this randomness is magnified by a system that provides a discontinuous and significant reward (i.e., bonus points) for making mission in a single month. A substantial number of stations

---

[4] Recently, USAREC began implementing a battalion mission-allocation model based on past enlistments in the Army and other services. We believe that this approach has merit and is likely to improve missioning, at least at the battalion level. Such a model, however, is not used to allocate the battalion mission to the station level, nor is it flexible enough to adjust adequately to relative changes in the local recruiting environment.

that often achieved regular Army mission are actually less, or no more, productive than many of their counterparts that make mission less frequently.

- Consider a more refined system of rewards for additional enlistment categories such as males, youth with more education or higher AFQT scores, youth with skills enabling them to fill critical MOSs, or those willing to accept longer terms of service.

Our exploratory research distinguishing males and females suggests that other market segments that we have not analyzed may also differ significantly in terms of recruiting difficulty. These differences are likely to vary systematically across the market areas of different stations. They could also vary when there are changes in the distribution of needed enlistments by MOS. It would not be advantageous to allocate missions for detailed subcategories, however, especially if mission accomplishment in a single month continues to lead to substantial bonus points. This is because doing so would increase the importance of randomness in performance evaluation and rewards. However, explicit recognition of market-quality differences in establishing recruiting goals or even a supplemental reward system providing points based on the distribution of enlistments among categories of differing importance to USAREC would better reward productivity as well as provide additional incentives for recruiters to help meet overall Army objectives.

- To minimize resistance, include education and outreach when implementing reforms.

Organizational change is always difficult, especially when there are perceived winners and losers. Modest efforts to explain and justify the changes could substantially increase their acceptance. If the performance measures are perceived as fair, with every station given a reasonable chance of success if their recruiters work hard, resistance will be reduced.

   Although it is impossible to quantify the productivity gains that could emerge from such reforms, they are likely to dwarf any costs of implementation. As demonstrated in previous research (Dertouzos and Garber, 2006), better mission allocations during 2001 to 2003 could have improved average recruiting productivity for high-quality enlistees by nearly 3 percent. Much of this gain would have been due to an increased willingness on the part of stations that had a recent history of success (by conventional measures) to work harder and be more responsive to mission increases. There is substantial reason to believe that using a performance metric that better reflects Army values and more accurately assesses recruiter effort and skill would also have significant benefits.

# Acknowledgments

We are thankful to several people who were generous with their advice, comments, and information. In particular, we are grateful to Rod Lunger, from the United States Army Recruiting Command, who provided much of the data used in this analysis. Jan Hanley and Stephanie Williamson of RAND provided extensive expert assistance in obtaining and processing data. Martha Friese and Nancy Good helped prepare the manuscript. We are extremely thankful to Bruce Orvis for his valuable input and support from the inception of this project. Lastly, Jim Hosek and John Romley of RAND provided careful, constructive and insightful technical reviews that helped us improve this report in a variety of ways. It goes without saying that opinions expressed and any remaining errors are the responsibility of the authors.

# Abbreviations

| | |
|---|---|
| AFQT | Armed Force Qualification Test |
| BOX | Regular Army Mission Box |
| DEP | Delayed Entry Program |
| FY | fiscal year |
| HM | high-quality mission |
| HQ | high quality |
| HWR | high-quality write rate |
| MOS | Military Occupational Specialty |
| OPRA | on-production regular Army |
| PPM | preferred performance metric |
| QMA | qualified military available |
| TM | total mission |
| TWR | total write rate |
| USAR | U.S. Army Reserve |
| USAREC | U.S. Army Recruiting Command |

# CHAPTER ONE

# Introduction

The United States Army has several human resource policies at its disposal to enhance the productivity of its recruiting force. Such policies include recruiter selection and assignment, setting enlistment goals, and rewarding successful recruiters. Recent RAND research by the present authors analyzed prevailing personnel policies and concluded that, during the period from June 2001 through September 2003, the Army could have increased recruiter productivity at little or no cost by implementing modest changes in these practices.[1]

The present study focuses on measurement and assessment of recruiting performance. Performance metrics are important because they are the standard by which individuals and organizations are judged. Thus, they can motivate individuals and influence their choices; in the context of Army recruiting, recruiter choices can have major impacts on the ability of the Army to meet its goals. Clearly, the Army's performance metrics should be designed and used to support Army goals. Doing so requires analysis of how different metrics would affect recruiter behavior and, in turn, recruiters' contributions to achieving the Army's goals. To induce maximum effort, the Army's performance metrics must be viewed by recruiters as sufficiently fair so as not to undermine recruiter morale.[2]

---

[1]  See Dertouzos and Garber (2006).

[2]  For some discussion about performance metrics and targets, see Chowdhury (1993, pp. 28–41) and Darmon (1997, pp. 1–16).

This research extends our earlier study in several ways. First, we analyze data for a longer period, namely, fiscal years (FYs) 2001–2004. Second, we extend the econometric models and analyses of our earlier report. Third, we develop a conceptually grounded performance metric—which we call the *preferred performance metric* (PPM)—to estimate the effort and skill applied by a station's recruiters to produce contracts of various types. This metric is "preferred" because using it would provide incentives for (1) current recruiters to "work hard and work smart" and (2) soldiers who have good skills or aptitudes for recruiting to volunteer for recruiting duty. Fourth, we compare performance metrics currently used by various organizational layers of the United States Army Recruiting Command (USAREC) with three alternative PPMs.

Common or *traditional* performance measures, which can be computed for command or organizational units of various sizes (e.g., stations, companies, or battalions) and time intervals of varying lengths, include the following:

- How high was the total write rate (TWR) per recruiter? That is, how many enlistment contracts were signed on average?
- How many high-quality prospects (i.e., high school seniors and graduates scoring in the top half of the Armed Forces Qualification Test) were signed per recruiter?[3]
- By what percentage did the command unit exceed or fall short of targets (or missions) for total and for high-quality enlistments?
- How often did the command unit make regular Army mission box?

Ideally, a performance metric would reflect recruiter and commander abilities and effort levels while controlling for such exogenous factors as the quality of the unit's market territory and changes in enlistment propensity. To accurately reflect effort levels, performance

---

[3]  Throughout this report, we use the terms *high-quality* and *alpha* as they are used within USAREC, to refer to high-school seniors and high-school graduates scoring in the top half (equivalently, in categories I through IIIA) of the Armed Forces Qualification Test (AFQT). So, for example, the terms "high-aptitude seniors," "high-quality seniors," and "senior alphas" are synonymous.

metrics must account for the relative difficulty of recruiting different categories of youth (e.g., high-quality seniors versus high-quality graduates). Finally, measures should not be unduly influenced by random factors affecting enlistment outcomes. In other words, merely being "luckier" should not often be the basis for recruiting units to be designated as higher performers.

In Chapter Two, we generalize a contract-production model developed in Dertouzos and Garber (2006) and develop our PPM conceptually. This conceptual development requires specification and analysis of a microeconomic model of effort allocation by recruiters; that model and analysis are detailed in Appendix A. The PPM explicitly recognizes that there are multiple enlistment categories and variations in local market quality that affect the difficulty of recruiting within these separate categories. In Chapter Three, we present estimates of econometric models of recruiting outcomes using monthly, station-level data for FYs 2001 to 2004. We first estimate a model distinguishing market quality for the three categories of youth that are missioned separately during this period: high-quality, high school graduates; high-quality, high school seniors; and other enlistments. These estimates provide the empirical foundation for comparing alternative performance measures. We conclude Chapter Three with an exploratory analysis of four enlistment categories that are not missioned separately, namely, contracts broken down by quality and gender. In Chapter Four, we compute three versions of our PPM and compare them with five traditional performance measures. In Chapter Five, we consider two additional issues related to the design of performance metrics. First, we evaluate alternative *performance windows* or time periods, such as month, quarter, or year. Second, we evaluate the efficacy of missioning and performance evaluation for the individual recruiter versus the station. Chapter Six concludes with a discussion of the policy implications of our research findings.

CHAPTER TWO

# Models of Recruiter Effort, Market Quality, and Enlistment Supply

In this chapter, we present the econometric models we employed in our empirical work and recruiter performance measures based on those models. For given recruiting stations in particular months, the models relate enlistment outcomes (contracts signed) to recruiter effort and the quality of the recruiting market. We begin by reviewing a model used by Dertouzos and Garber (2006, Chapter Four) that focuses on a single enlistment outcome—namely, contracts signed by high-quality recruits.[1] This review provides background and context for a new model that distinguishes among the three contract types that are directly missioned, which we present subsequently.

The key ideas underlying these models are the following:

- The level of effort expended by recruiters depends on the difficulty of achieving their enlistment goals, their recruiting skills,[2]

---

[1] Dertouzos and Garber (2006) provide additional motivation for this approach (e.g., from the literature on managing sales forces in psychology and management) and additional caveats.

[2] The role of recruiter skill (or, synonymously, talent or aptitude) in producing contracts is not explicitly discussed in Dertouzos and Garber (2006) because distinguishing level of effort from level of skill was not important for the purpose of that report, which was to estimate the relative quality of different station areas or markets for recruiting high-quality youth and to determine implications for missioning. The focus of this report is performance measurement; thus, the role of recruiter skill, which complements effort in producing contracts, is important. We make the role of skill explicit later in this chapter when we present our approach for studying more than one contract type.

5

and their past success as recruiters.[3]

- The difficulty of achieving an enlistment goal depends on the goal and the quality of the market area assigned to the recruiter's station.
- When the difficulty of achieving an enlistment goal is low, increasing the difficulty will increase effort.
- If, however, the difficulty of achieving an enlistment goal is sufficiently high, increasing the difficulty *may* decrease effort.
- The expected number of enlistments for a station in a particular month depends on the quality of the market, the total effort expended by the station's on-production recruiters, and the average skill level of those recruiters.

## A Model with a Single Type of Contract

In Dertouzos and Garber (2006), we used monthly station-level data from January 2001 through June 2003 to develop and estimate a model focusing on high-quality enlistees that involved estimation of market quality for each station in each month. In the next section, we generalize that model to consider determination of station-level production of the three contract types that have separate missions.[4] Before we present the generalized model and estimates based on it, we briefly review the model developed earlier. Let the subscript $s$ denote a recruiting station. For each recruiting station and month,[5] let

$c_s$ = high-quality (HQ) contracts signed (in station $s$ in a particular month)

$m_s$ = high-quality mission

---

[3]   The role of past productivity could be based on several factors, including an improvement in morale, increased confidence that recruiters can make current performance targets, or an increase in promotion incentives due to the higher likelihood of cumulative success.

[4]   These three types are high-quality (high-school) graduates (*grad alphas*), high-quality (high-school) seniors (*senior alphas*) and "others."

[5]   For economy of notation, we suppress the month index throughout this chapter.

$l_s$ = high-quality Delayed Entry Program (DEP) losses (charged that month)

$g_s \equiv m_s + l_s$ = high-quality enlistment goal[6]

$N_s$ = number of on-production regular Army (OPRA) recruiters

$e_{is}$ = effort level of OPRA recruiter $i$ in station $s$ $(i = 1, 2, ... N_s)$

$e_s = N_s e_{is}$ = total effort by all OPRA recruiters in the station[7]

$c_s^*$ = the marginal productivity of recruiter effort, which depends on the quality of the recruiting market in the station's territory and other factors.

We conceptualize the quality of a station's market as a component of the marginal product of recruiter effort ($c_s^*$) in producing high-quality contracts in its recruiting territory (i.e., recruiter effort is more productive, other things equal, in better markets). In particular, we assume that the expected number of high-quality contracts signed by a station is given by

$$Ec_s = c_s^* e_s. \tag{2.1}$$

Thus, high-quality contracts increase with both effort and market quality, which are mutually reinforcing.[8] Rearranging (2.1) yields

$$e_s = Ec_s / c_s^*, \tag{2.2}$$

---

[6]  As reported in Dertouzos and Garber (2006, p. 34), in order to succeed with regard to regular Army recruiting in a particular month during our sample period, a station's recruiters must meet (or exceed) for each missioned category of recruits (i.e., grad alphas, senior alphas and others) the station's mission including any DEP losses charged that month. Rules in effect during our sample period typically allowed grad alphas to substitute for senior alphas and either type of contract to substitute for "other" in determining whether a station made mission.

[7]  We assume that in a given month every OPRA recruiter in a station expends the same level of effort.

[8]  Equation (2.1) implies that the marginal productivity of effort is constant. It is worth noting, however, that the units of measurement for effort are defined in terms of the expansion in enlistments, holding market quality constant. The actual but unobserved activities required to achieve constant increments in contracts may, in fact, involve increases in the amount or intensity of time actually spent.

which shows that the effort required from all recruiters in a station to achieve a given level of expected contracts is inversely proportional to $c_s^*$. It proves useful to interpret $1/c_s^*$ (the inverse or reciprocal of the marginal productivity of effort) as the difficulty of recruiting high-quality youths in the market territory of station $s$.

Despite the fact that during the sample period missions were assigned at the station level, it is helpful to formalize the concept of an average recruiter's difficulty in meeting his or her share of the station's high-quality goal. To do so, consider a station with $N_s$ recruiters on production, a monthly goal equal to $g_s$, and marginal productivity of effort equal to $c_s^*$. For that station's *expected* contracts to equal $g_s$, equation (2.2) implies that total effort by all OPRA recruiters must be $e_s = g_s/c_s^*$, and the effort required per recruiter is then $e_{is} = g_s/(N_s c_s^*)$. Accordingly, we define the difficulty of making recruiter $i$'s share of the goal of (his or her) station $s$ as

$$d_{is} = g_s/(N_s c_s^*),\qquad(2.3)$$

which (according to the previous sentence) is also the expected level of effort required of an average on-production regular Army recruiter to achieve the monthly station goal, given the quality of the station's market area and other factors that determine the marginal productivity of effort.

Equation (2.1) relates expected high-quality contracts to effort and the marginal productivity of effort, neither of which is measurable (i.e., they are *unobservable*). For estimation purposes, we relate these unobservable concepts to observable variables as follows.

First, we assume that effort per recruiter in a station in any month is the same for all recruiters within the station and that effort depends on the difficulty faced by each recruiter as well as the station's recent past success in enlisting high-quality youth. Specifically, total effort by all recruiters in station $s$ is assumed to be determined by

$$e_s = N_s[1 + (\beta + \beta_1 R_s)d_{is} + (\gamma + \gamma_1 R_s)(d_{is})^2]\qquad(2.4)$$

where effort per recruiter is given by

$$e_{is} = 1 + (\beta + \beta_1 R_s)d_{is} + (\gamma + \gamma_1 R_s)(d_{is})^2 \qquad (2.4')$$

and $R_s$ equals the ratio of the station's high-quality enlistments to its high-quality mission over the twelve-month period ending three months before the current month. This ratio is a measure of recruiting success of station $s$ in the recent past, and is included in the effort equation because (as discussed in Dertouzos and Garber, 2006, p. 73) the literature in psychology and sales-force management indicates that people tend to exert more effort in performing tasks if, other things being equal, they have more confidence in their abilities to perform.[9]

The effort equation (2.4) also incorporates the assumption that the level of recruiter effort depends on the difficulty of station $s$ making the regular Army mission for high-quality youth, as defined in (2.3). The difficulty of making mission enters (2.4) in both its level and its square—i.e., the relationship between effort and mission difficulty is assumed to be quadratic or parabolic—to allow for the possibility (but not to impose a restriction on the estimates) that when the difficulty of making mission is high, effort could fall in response to an increase in difficulty resulting from an increase in mission.[10]

The assumption in (2.4') that the intercept in the equation for effort-per-recruiter equals 1 is a normalization that serves to specify or pin down the scales on which effort and the marginal productivity of effort are implicitly measured. This normalization is innocuous in the present case of a model involving only a single contract type.[11]

---

[9] In fact, Dertouzos and Garber (2006, pp. 80–83) found empirically that recruiting success in the recent past is an important determinant of recruiter effort.

[10] See Dertouzos and Garber (2006, pp. 62–63) for further discussion. In fact, their estimates (which they report in Table 4.1, pp. 77–79) are consistent with the hypothesis that effort increases at a decreasing rate—and eventually declines—as the difficulty of making mission increases. But Dertouzos and Garber (2006, pp. 80–81) also conclude from their estimates that during their sample period (January 2001 through June 2003) virtually no recruiting stations faced difficulty great enough to imply that their recruiters would have put forth more effort if their missions had been lower.

[11] See Dertouzos and Garber (2006, pp. 73–74). As we discuss below, however, it is much more challenging to pin down the scales of effort and the marginal productivity of effort in generalizations involving more than one contract type in a way that results in the levels of

Second, regarding $c_s^*$ (which is also unobservable) we assume that the marginal (and average) product of recruiter effort for a station in a particular month is linearly related to observable variables—some of which control for the quality of the market and others of which control for other factors affecting the marginal productivity of effort, such as staffing, features of the station's Army reserve recruiting, and seasonal factors—contained in the vector $x$:

$$c_s^* = \alpha' x_s . \tag{2.5}$$

To derive the form of the nonlinear regression equation estimated by Dertouzos and Garber (2006, Chapter Four), let $y_s$ denote monthly high-quality contracts produced by station $s$ (the dependent variable in our regressions). Combining equations (2.1), (2.3), (2.4), and (2.5) yields the following expression for expected high-quality contracts signed by a station's recruiters in a particular month:

$$
\begin{aligned}
Ey_s = Ec_s = c_s^* N_s e_{is} &= N_s c_s^* [1 + (\beta + \beta_1 R_s)(g_s / N_s c_s^*) \\
&\quad + (\gamma + \gamma_1 R_s)(g_s / N_s c_s^*)^2] \\
&= N_s c_s^* + (\beta + \beta_1 R_s) g_s + (\gamma + \gamma_1 R_s)(g_s^2 / N_s c_s^*) \tag{2.6}
\end{aligned}
$$

where $c_s^* = \alpha' x_s$ .

## A Model Distinguishing the Three Missioned Contract Types

In this report, we generalize the model just described to (1) make explicit the role of recruiter skill in producing enlistments and (2) distinguish among the three categories of enlistees that are missioned by USAREC: (1) high-quality, high school graduates (*grad alphas*),

---

effort directed toward enlistments of different types all to be measured on the same scale (e.g., hours of standardized-quality effort).

(2) high-quality, high school seniors (*senior alphas*) and (3) all other enlistees (*others*).

We generalize the earlier model for several reasons. First, the quality of a recruiting station's territory for enlisting one type of prospect (e.g., senior alphas) may not be very informative about the quality of that territory for enlisting youths of the other two types. Second, an enhanced understanding of which station areas are, for example, better for signing seniors versus graduates could support development of missioning models that would increase recruiter productivity. Third, performance measures based on contracts of different types should account for the variations across geographic areas and contract types in the difficulty of recruiting different categories of youth. Finally, a fundamental issue for our purposes is whether, in interpreting contracts produced for the purposes of performance assessment, the effects of effort and skill can be disentangled and—in any event—whether performance measures should reward recruiter effort, skill, or both.

Let the subscript $j$ denote the contract type, which can take on three values: $G$ (for grad alphas), $S$ (for senior alphas), and $O$ (for others). Generalizing the notation of the previous section, let

$c_{sj}$ = contracts of type $j$ signed (in station $s$ in a particular month), for $j = G, S, O$,

$m_{sj}$ = mission for contract type $j$

$l_{sj}$ = DEP loss of type $j$ (charged in that month)

$g_{sj} \equiv m_{sj} + l_{sj}$ = enlistment goal for type $j$

$e_{isj}$ = effort level of OPRA recruiter $i$ ($i = 1, 2, ... N_s$) directed toward signing youth of type $j$

$e_{sj} = N_s e_{isj}$ = total effort by all OPRA recruiters in the station directed at contract type $j$[12]

$v_{sj}$ = total (across recruiters) skill level of recruiters in station $s$ for signing youth of type $j$

---

$c_{sj}^{*} =$ marginal productivity of recruiter effort directed at enlisting youths of type $j$ in producing type-$j$ contracts, which depends on the quality of the market in the territory of station $s$ for recruiting youths of type $j$ and other factors.

The generalization of the contract production equation (2.1) for our three-contract-type model is:

$$Ec_{sj} = c_{sj}^{*}(e_{sj} + v_{sj}) \text{ for } j = G, S, O. \qquad (2.7)$$

Implicit in the three equations stated in (2.7) are the assumptions that (1) holding recruiter effort constant, expected contract levels rise as recruiter skill grows, (2) recruiter effort can be directed at particular contract types, and only effort directed at a particular contract type can produce contracts of that type, (3) markets that are relatively good for recruiting youth of one type may or may not be good for recruiting youths of the other two types (because, for any particular station, $c_{sj}^{*}$ may vary considerably across contract types), and (4) recruiting skill varies across stations and across contract types within stations.[13]

Note that all three variables on the right-hand side of equation (2.7)—namely, effort and skill levels and the marginal productivity of their sums—are unobservable. Moreover, the scales on which these variables are measured are not uniquely determined by the scales on which the observable variables (our data) are measured. To see this, observe from equation (2.7) that the value of expected contracts of type $j$ (i.e., $Ec_{sj}$) would not change if we were to change the scale on which (or, equivalently, change the units in which) $c_{sj}^{*}$ is measured by dividing its values by a constant while at the same time rescaling $e_{sj}$ and $v_{sj}$ by multiplying their values by the same constant. Thus, either set of scales for these three sets of variables (and an infinite number of others) has the same implications for observable outcomes of these variables (i.e., numbers of contracts produced). All such sets of scales

---

[13] Among these assumptions, the second stands out as being particularly strong. Its empirical implications—when combined with our assumptions about the determinants of recruiter effort in equation (2.8)—are discussed below.

are, then, observationally equivalent, and without further restrictions on the parameters of the empirical model, these parameters cannot be estimated (i.e., they are *not identified*).

As described in the previous section, there is an analogous ambiguity about the scales of effort and the marginal productivity of effort in the case of the model of production of contracts of a single type, and this ambiguity was resolved in Dertouzos and Garber (2006) by a normalization that arbitrarily set (or pinned down) the scales of those two unobservable variables. In that case, the normalization took the form of assuming a particular value (i.e., 1) for the intercept of the equation determining effort per recruiter (in (2.4')). For our generalized model and empirical analysis involving more than one contract type, however, arbitrary rescaling—e.g., by assuming intercepts of 1 for all three effort-per-recruiter equations—will not suffice for the purposes of performance measurement. For example, suppose that for performance assessment we sought to estimate total recruiter effort across contract types for a particular station-month. To sum the estimated effort levels across contract types, these three effort levels must be measured on a common scale (i.e., in the same units)—for example, hours of standard-intensity time spent in recruiting activities. But choosing arbitrary scales for the three effort levels will result in these variables being measured in different *unknown* units (e.g., hours versus half-hours of standard-intensity time), in which case they cannot be meaningfully summed. In short, summing quantities of effort that are measured in different units would be akin to adding apples and oranges.

To construct conceptually grounded performance measures, we need to restrict the parameters of the generalized model in a way that implicitly measures effort and skill using the same units across contract types.[14] As detailed in Appendix A, to resolve this problem for our

---

[14] In view of (2.7), effort and skill directed at enlistments of any given type must be in the same units because effort and skill are summed there. Note that if effort and skill are measured in the same units across contract types, then this will also be true of the marginal productivities of effort and skill in producing contracts of different types. (To see this, note from (2.7) that the observable contract levels are all measured in enlistees per month, and that expected contract levels are equal to the products of marginal productivities and effort plus skill.)

model with three types of enlistments, we employ a microeconomic analysis. More specifically, in our microeconomic model recruiters choose optimal (*utility-maximizing*) levels of effort to direct toward recruiting youth of each of the three types. These optimal levels of effort balance recruiters' (tangible and intangible) rewards (*utility*) from producing additional contracts against the negative consequences (*disutility*) of expending additional effort. This analysis yields explicit solutions for the three effort-per-recruiter equations (analogs to (2.4′)) and, most important, shows that the relative values of the intercepts of the equations determining these three effort levels equal known constants. Imposing these relative values in estimation determines the *common* (or same-unit) scales of the three effort (and three corresponding skill) levels. Having thus pinned down the scales *in the same units*, we can then sum effort and skill levels across contract types. In our preferred, or *base-case*, interpretation, these relative intercepts are, in fact, the points awarded by USAREC to stations and their recruiters for signing youth of the three types.[15]

Relying on this microeconomic analysis, the equations determining station-level effort per recruiter directed at each contract type are specified with known intercepts $\pi_j$ (i.e., these values are imposed in estimation) of our three effort-per-recruiter equations

$$e_{isj} = \pi_j + (\beta_j + \beta_{jj}R_{sj})d_{sj} + (\gamma_j + \gamma_{jj}R_{sj})d_{sj}^2 \text{ for } j = G, S, O, \quad (2.8)$$

where: (a) $d_{sj} = g_{sj} / N_s c_{sj}^*$, which generalizes (2.3), is the difficulty for a recruiter in station $s$ to achieve his or her share of the station's monthly goal for contracts of type $j$; (b) $R_{sj}$, which generalizes $R_s$ in the one-contract-type model presented in the previous section, is a

---

[15]  We also consider two other sets of the relative value to recruiters of contracts of different types that span a wide range of possibilities—arguably, the full plausible range. At one extreme, we assume that recruiters value all types of contracts equally, in which case the relative values are all equal to 1. At the other extreme, we assume that USAREC values only high-quality graduates and seniors. Our preferred assumption—embodied in PPM2 (see below)—is that recruiters value contracts of different types in proportion to points awarded because a key purpose of assigning different point values to different types of recruits (along with separate missions for the different types) is to communicate to the field USAREC's priorities concerning contract types.

measure of the station's recent past success in recruiting youths of type $j$;[16] and (c) $\beta_j, \beta_{jj}, \gamma_j, \gamma_{jj}$ are parameters to be estimated. Then our estimating equations—the analogs to (2.6)—which are also nonlinear in the parameters, are, for $j = G, S, O$

$$Ey_{sj} = Ec_{sj} = c_{sj}^* N_s e_{isj} = N_s c_{sj}^* [\pi_j + (\beta_j + \beta_{jj} R_{sj})(g_{sj} / N_s c_{sj}^*)$$
$$+ (\gamma_j + \gamma_{jj} R_{sj})(g_{sj} / N_s c_{sj}^*)^2]$$
$$= N_s c_{sj}^* \pi_j + (\beta_j + \beta_{jj} R_{sj}) g_{sj} + (\gamma_j + \gamma_{jj} R_{sj})(g_{sj}^2 / N_s c_{sj}^*)(2.9)$$

where $c_{sj}^* = \alpha_j' x_s$ and the $\alpha_j$ are type-specific parameters to be estimated along with $\beta_j, \beta_{jj}, \gamma_j$ and $\gamma_{jj}$. Thus, in estimating the three-contract-type model we use the same observable predictors (i.e., the variables contained in the vector of variables $x_s$ do not vary over $j$) of the marginal productivities of effort and skill, and we allow these variables to have different coefficients (i.e., the values of the elements of the vector of parameters $\alpha_j$) in the equations for the different contract types.

Two strong assumptions are implicit in equation (2.9): (1) only effort directed at a particular contract type can produce contracts of that type (as mentioned in the context of equation (2.7)), thus ruling out spillovers in production, and (2) effort directed at one type of enlistment does not depend on goals for other types, which is implicit in (2.8). Jointly, these assumptions imply that contracts for one type do not depend on goals for other types, which simplifies the empirical model considerably. This proposition, however, is conceptually dubious because, if a higher goal for one contract type increases effort directed at that type (as implied by (2.7)), one might reasonably expect that this higher goal also tends to reduce the levels of effort to recruit youth of the other two types. We investigated empirically the implication of our model that effort levels do not depend on goals for other types of contracts by allowing for cross-contract-type effects of goals on effort levels in our estimating equations. In fact, we found positive cross-type

---

[16] Specifically, $R_{sj}$ is the ratio of the station's enlistments of type j to its mission for type j over the twelve-month period ending three months before the current month.

effects, suggesting positive spillovers in production, effort, or both. However, including these cross-type effects in our empirical analyses did not substantially affect our policy conclusions and, hence, are not reported here.

## A Conceptually Grounded, Econometrically Based Performance Measure

A key purpose of performance measurement is to provide incentives for changing behavior in ways that increase enlistments or, equivalently, recruiter productivity. In the short term, the role of such incentives is to induce recruiters to work harder and smarter. In the longer term, performance measurement should encourage soldiers with relatively good skills for recruiting, or who can develop such skills in training for recruiting, to volunteer for recruiting duty, because soldiers who are more skilled at recruiting produce more contracts for a given level of effort, holding market quality constant.[17] Thus, USAREC might best use a performance measure that provides both types of incentives—a measure that would involve rewarding both effort and skill. In fact, available empirical information does not allow us to separate the contributions of recruiter effort and recruiter skill in producing enlistment contracts. Thus, we cannot construct performance metrics that are reasonably interpreted as reflecting effort alone even if USAREC would prefer to focus performance measurement entirely on assessment of effort. This inability can be understood intuitively by noting that the only empirical information about recruiter skill in our model consists of enlistment outcomes and exogenous factors affecting the

---

[17] In our microeconomic analysis in Appendix A, we show that, other things being equal, more highly skilled recruiters will expend less effort, but that effort plus skill—and, hence, productivity (see (2.7))—will be higher for more highly skilled recruiters. Thus, even though better recruiters may benefit personally in the form of needing to exert less effort to meet their goals, it is to USAREC's advantage to use performance measurement to encourage soldiers who have good sales aptitudes or skills (holding constant the opportunity cost to the Army of taking these soldiers out of their primary military occupational specialties [MOSs]) to volunteer for recruiting.

difficulty of recruiting. Thus, since skill and effort enter the model in the same way in terms of their effects on enlistments—see (2.7)—and we cannot measure or estimate effort or skill directly, we cannot infer how much contract production results from skill and how much results from effort.[18]

In Appendix A we derive a station-level performance measure (given by (A.6)) that combines, in a particular way that requires econometric estimation of marginal productivities of effort, contracts of each of the three types produced per OPRA recruiter. This measure, which explicitly takes into account the difficulty of recruiting youth of different types, estimates the sum across contract types of the *effort plus skill* applied to recruiting each type of youth given the three observed levels of contract production. This measure for station *s* during some period (number of months) is our *preferred performance metric* (PPM):

$$PPM = \frac{1}{N_s}[\frac{c_sG}{c_{sG}^*} + \frac{c_{sS}}{c_{sS}^*} + \frac{c_{sO}}{c_{sO}^*}] , \qquad (2.10)$$

where, as previously defined, (a) $N_s$ is the number of OPRA recruiters in station *s*, (b) the $c_{sj}$ (for $j$ = G, S, O) are numbers of contracts produced by station *j* during the period considered for performance evaluation, and (c) the $c_{sj}^*$ connote the unobservable, but estimable, marginal productivity of effort in producing enlistments of the three types. To implement this measure, we need data on numbers of OPRA recruiters and contracts produced by type and estimates of the $c_{sj}^*$ .

The intuitive appeal of this measure can be seen as follows. Recall that $1/c_{sj}^*$ is the difficulty of recruiting youths of type *j* in the market territory of station *s*. Thus, the metric given by (2.10) is the sum over contract types of the numbers of contracts produced, each multiplied (or weighted) by the difficulty of producing that contract type in this

---

[18] We do have empirical information that is helpful in predicting effort levels, which are endogenous in—i.e., determined by—our model, namely, missions and the determinants or predictors of marginal productivity of effort. But, lacking data that would be useful for predicting or controlling for variations in skill levels, which are exogenous in our model, we implicitly incorporate the effects of skill on enlistments in the error or disturbance terms appended to equations (2.9) for purposes of estimation.

market area, expressed on a per-OPRA-recruiter basis. As discussed in the previous section, different assumptions about the relative values recruiters assign to contracts of different types (as embedded in their utility functions) lead to different restrictions on the intercepts (i.e., the $\pi_j$) of the effort-per-recruiter equations (2.8), which in turn produce different sets of estimates of the $c_{sj}^*$ from estimation of (2.9) with different constrained values for these parameters and different sets of numerical values of the PPM in (2.10). We refer to different sets of estimates of the reciprocals of $c_{sj}^*$ (which are estimates of the difficulty of recruiting youth of different types) as different weights because—as can be seen by inspecting (2.10)—they assign different levels of importance to contracts of different types. And we refer to different sets of estimates of the PPM resulting from different sets of weights as different versions of the PPM (which we call PPM1, PPM2 and PPM3 in Chapter Four.)

We conclude this section by pointing out another intuitively appealing property of the PPM, namely, that it implicitly gives more weight to contract types that are more highly valued by recruiters—presumably, because they are more highly valued by their leaders who communicate their preferences in various ways. Although the expression for the PPM does not *explicitly* involve the values recruiters assign to contracts of different types in their utility functions, these values are implicit in the PPM formula. This can be seen heuristically as follows. As we show in the analysis detailed in Appendix A, the values recruiters assign to different contract types in their utility functions—which are denoted as $\pi_j$ in the expression for their utility functions detailed in equation (A.1)—are the same values used to constrain the intercepts of the effort-per-recruiter equations (2.8). And, as suggested by (2.7), a higher value for the intercept of an effort-per-recruiter equation corresponds to a smaller value for the corresponding $c_{sj}^*$, because the values and scales of the observed contract levels ($c_{sj}$) are fixed (at, for example, two grad alpha contracts per month). For example, a doubling of the intercept for grad alphas will reduce the size of the implicit $c_{sG}^*$ by half (and tend to do the same to its estimated values). Thus higher relative values for an intercept in an effort-per-recruiter equation will tend to produce higher weights (the reciprocals of the estimated $c_{sj}^{*}$) in computing PPMs, which makes good sense intuitively. In sum, more highly valued enlistment categories receive more weight in our PPM.

# CHAPTER THREE

# Data and Econometric Estimates of Contract-Production Models

In this chapter, we describe our data and then present estimates of the parameters of the model presented in Chapter Two that distinguish among the three types of separately missioned enlistments. We then present a model that distinguishes four categories of enlistments that are not missioned separately: (a) high-quality men, (b) high-quality women, (c) other men, and (d) other women—and present the estimates for that model. The latter model exemplifies how effort and market-quality levels might be distinguished for enlistment categories that are not separately missioned.

## Data

To estimate the two models, we used data from more than 1,500 stations observed monthly during FYs 2001–2004. Table 3.1 defines variables and reports sample means and standard deviations.

The enlistment variables include total numbers of AFQT I-IIIA (high-quality) graduates and seniors, as well as all other contracts. The average station signed just over two graduates per month. For the average station of 2.48 recruiters, this average production level represents about 0.8 high-quality graduate enlistments per recruiter per month. Many fewer high-quality high school seniors enlist before graduating, with seniors representing about 27 percent of the high-quality pool. Enlistments by nongraduates and those in lower AFQT categories

**Table 3.1**
**Monthly Station-Level Data, FYs 2001–2004**

| Variable | Description | Mean | Standard Deviation |
|---|---|---|---|
| **Dependent variables** | | | |
| High-quality men | Male high-quality contracts (AFQT I-IIIA graduates and seniors) | 2.1790 | 1.7071 |
| High-quality women | Female high-quality contracts (AFQT I-IIIA graduates and seniors) | 0.5600 | 0.8820 |
| Other men | All other male contracts | 1.4608 | 1.3685 |
| Other women | All other female contracts | 0.4028 | 0.7067 |
| High-quality graduates | High school graduate contracts (AFQT I-IIIA, male and female) | 2.0392 | 1.7322 |
| High-quality seniors | High school senior contracts (AFQT I-IIIA, male and female) | 0.7555 | 0.9411 |
| **Regular Army (RA) mission variables** | | | |
| Senior mission plus DEP loss | Senior mission (AFQT I-IIIA) plus DEP losses (male and female) | 1.0528 | 0.8615 |
| Graduate mission plus DEP loss | Graduate mission (AFQT I-IIIA) plus DEP losses (male and female) | 2.5905 | 1.5297 |
| Other mission plus DEP loss | Other mission, plus DEP losses (male and female) | 2.0331 | 1.3886 |
| **MOS availability variables** | | | |
| Combat support | Percent of national enlistments in combat-support MOSs | 0.1977 | 0.0217 |
| White collar | Percent of national enlistments in white-collar MOSs | 0.2498 | 0.0352 |
| Blue collar | Percent of national enlistments in blue-collar MOSs | 0.1880 | 0.0156 |
| Combat | Percent of national enlistments in combat MOSs | 0.3644 | 0.0448 |
| **Recent past success variables** | | | |
| Senior ratio | Ratio of senior production to mission for previous year, lagged 3 months | 0.7260 | 0.3874 |
| Graduate ratio | Ratio of graduate production to mission for previous year, lagged 3 months | 0.7748 | 0.2949 |
| Other ratio | Ratio of "other" production to mission for previous year, lagged 3 months | 1.0278 | 0.3673 |

## Table 3.1—continued

| Variable | Description | Mean | Standard Deviation |
|---|---|---|---|
| **Recruiter variables** | | | |
| Recruiters | Regular Army recruiters on production | 2.4846 | 1.2360 |
| 2-recruiter station | Dichotomous variable = 1 when there are 2 recruiters on production | 0.3332 | 0.4714 |
| 3-recruiter station | Dichotomous variable = 1 when there are 3 recruiters on production | 0.2336 | 0.4232 |
| 4-recruiter station | Dichotomous variable = 1 when there are 4 recruiters on production | 0.1298 | 0.3361 |
| 5-recruiter station | Dichotomous variable = 1 when there are 5 recruiters on production | 0.0512 | 0.2205 |
| 6+ recruiter station | Dichotomous variable = 1 when there are 6 or more recruiters on production | 0.0171 | 0.1295 |
| **Personnel status variables:** | | | |
| Commanders, on production | On production commanders, divided by total number of on-production recruiters | 0.2151 | 0.3313 |
| Recruiters on duty | Recruiters on duty, not assigned to production, divided by total on-production recruiters | 0.1082 | 0.2480 |
| Absent recruiters | Recruiters not on production, absent, divided by total on-production recruiters | 0.1234 | 0.3339 |
| Commanders, not on production | Commanders not on production, divided by total number of on-production recruiters | 0.1220 | 0.2565 |
| **Reserve variables:** | | | |
| Reserve recruiters | Reserve recruiters divided by number of regular Army on-production recruiters | 0.2188 | 0.3235 |
| Reserve mission, "other" | Reserve mission, "other," divided by number of regular Army on-production recruiters | 0.1716 | 0.3020 |
| Reserve mission, prior service | Reserve mission, prior service, divided by number of regular Army on-production recruiters | 0.2125 | 0.4398 |
| Reserve mission, high school | Reserve mission, high school, divided by number of regular Army on-production recruiters | 0.0910 | 0.1054 |
| DEP loss, "other" reserves | DEP loss, "other" reserves, divided by number of regular Army on-production recruiters | 0.0312 | 0.1531 |

## Table 3.1—continued

| Variable | Description | Mean | Standard Deviation |
|---|---|---|---|
| DEP loss, prior service reserves | DEP loss, prior service reserves, divided by number of regular Army on-production recruiters | 0.0007 | 0.0193 |
| DEP loss, high school reserves | DEP loss, high school reserves, divided by number of regular Army on-production recruiters | 0.0531 | 0.2085 |
| **Month indicator variables:** | | | |
| February | Dichotomous variable = 1 for the month of February | 0.0832 | 0.2762 |
| March | Dichotomous variable = 1 for the month of March | 0.0839 | 0.2772 |
| April | Dichotomous variable = 1 for the month of April | 0.0839 | 0.2772 |
| May | Dichotomous variable = 1 for the month of May | 0.0837 | 0.2770 |
| June | Dichotomous variable = 1 for the month of June | 0.0844 | 0.2779 |
| July | Dichotomous variable = 1 for the month of July | 0.0845 | 0.2782 |
| August | Dichotomous variable = 1 for the month of August | 0.0844 | 0.2779 |
| September | Dichotomous variable = 1 for the month of September | 0.0843 | 0.2779 |
| October | Dichotomous variable = 1 for the month of October | 0.0827 | 0.2754 |
| November | Dichotomous variable = 1 for the month of November | 0.0817 | 0.2739 |
| December | Dichotomous variable = 1 for the month of December | 0.0812 | 0.2732 |
| **Region indicator variables:** | | | |
| Mountain | Dichotomous variable = 1 for stations located in Mountain states | 0.0717 | 0.2580 |
| North Central | Dichotomous variable = 1 for stations located in North Central states | 0.2401 | 0.4271 |
| South | Dichotomous variable = 1 for stations located in Southern states | 0.3835 | 0.4862 |
| Pacific | Dichotomous variable = 1 for stations located in Pacific Coast states | 0.1378 | 0.3447 |

## Table 3.1—continued

| Variable | Description | Mean | Standard Deviation |
|---|---|---|---|
| **Local climate variables** | | | |
| Hot | Average July temperature (.1 degrees) | 753.8326 | 79.4648 |
| Rain | July precipitation (.01 inches) | 344.9914 | 204.9923 |
| Humidity | July humidity (percent) | 57.7849 | 15.0594 |
| **Market variables:** | | | |
| QMA per recruiter | Qualified military available (QMA) population, per OPRA recruiter, in logarithms | 6.3249 | 0.6459 |
| Unemployment change | Change in unemployment rate since last month, in logarithms | 0.0054 | 0.1131 |
| Unemployment level | Unemployment rate, in logarithms | 1.6424 | 0.3698 |
| Relative wage | Manufacturing earnings, divided by E-4 monthly compensation, in logarithms | −4.6765 | 0.1501 |
| **Demographic variables:** | | | |
| African American | Ratio of African American men to total men | 0.1325 | 0.1474 |
| Hispanic | Ratio of Hispanic to total men | 0.1591 | 0.1898 |
| College | Percentage of 17–21-year-old male population in college | 43.1453 | 4.6463 |
| Urban population[a] | Ratio of urban (census population 50,000 or greater) to total population | 0.5469 | 0.3859 |
| Clustered population | Ratio of urban cluster (census population of 2,500 to 49,999) to total population | 0.1620 | 0.1846 |
| Growth in single parent homes | Ratio of single-parent households in 2000 to single-parent households in 1990 | 1.3724 | 0.2667 |
| Poverty | Ratio of children in poverty to total population | 0.0069 | 0.0043 |
| Catholic | Ratio of adult Catholic adherents to total population | 0.1973 | 0.1435 |
| Eastern Religion | Ratio of adult Eastern religion adherents to total population | 0.0039 | 0.0069 |
| Christian | Ratio of adult non-Catholic Christian adherents to total population | 0.2341 | 0.1288 |
| **Veteran population variables:** | | | |
| Vet32 | Ratio of veteran population aged 32 or under to male population (17–21) | 0.1810 | 0.0727 |

**Table 3.1—continued**

| Variable | Description | Mean | Standard Deviation |
|---|---|---|---|
| Vet33-42 | Ratio of veteran population between 33 and 42 to male population (17–21) | 0.3383 | 0.1532 |
| Vet43-55 | Ratio of veteran population between 43 and 55 to male population (17–21) | 0.7477 | 0.2660 |
| Vet56-65 | Ratio of veteran population between 56 and 65 to male population (17–21) | 0.5890 | 0.2001 |
| Vet65-72 | Ratio of veteran population between 65 and 72 to male population (17–21) | 0.4108 | 0.1377 |
| Vet73 | Ratio of veteran population 73 or older to male population (17–21) | 0.6400 | 0.2557 |
| **Competition:** | | | |
| Army Market Share | Regular Army contracts as a percentage of total DoD active duty contracts, 1999 | 34.5661 | 7.8885 |

[a] Both urbanized areas and urban clusters are defined as densely populated areas having more than 500 people per square mile.

average 1.86 per station monthly. In other words, lower-quality contracts represent about 40 percent of the total. Female enlistees account for about 21 percent of both high-quality and other enlistments.

The data set also includes information on missions plus DEP losses (i.e., recruiting goals), broken down by I-IIIA seniors and graduates and all other enlistment categories. On average, monthly goals exceeded enlistments achieved, suggesting that the typical station underperformed relative to its production targets. For example, the production ratios, defined as the ratio of production over the past year (lagged one quarter) divided by the mission, was less than 0.8 for both high-quality seniors and graduates, indicating 20 percent underproduction on average.[1] On average, stations were able to achieve their lower-quality contract targets.

Our analysis of contract production for men versus women— categories that are not separately missioned—explores the roles of

---

[1]  For these measures, the ratios of the average enlistments to average missions are more meaningful than the average ratios for individual months. For example, missions for specific categories may be zero in a given month, even though the station may produce enlistments.

Army job availability or the composition of USAREC's demand for youth to train in military occupational specialties (MOSs) of different types. For this analysis, data were gathered on the monthly, command-wide (or nationwide) distribution of contracts signed for four broad MOS categories: combat support, white collar, blue collar, and combat, which are defined as follows.[2] First, all occupations that were not available to women were designated as *combat* MOSs. On average, about 36 percent of all contracts flowed into these MOSs during FY 2001 to 2004.

Next, all other jobs that had no obvious private-sector counterpart, such as weapons repair and missile operators, were designated as *combat support*. These represented about 20 percent of all contracts during this period. Jobs that had private-sector counterparts were placed in the *blue collar* or *white collar* categories according to the nature of the job. Blue collar MOSs, which account for about 19 percent of contracts, included construction jobs, truck maintenance, and transportation jobs. Finally, about 25 percent of all contracts were for MOSs corresponding to office, service, or professional occupations, such as nursing, clerical, or accounting. The relative importance of different categories varies over time. For example, the standard deviations in these measures relative to their means suggest that the number of available jobs could vary by over 10 percent from month to month. To the extent that variations in MOS distributions differentially affect the real or perceived training and career opportunities of men versus women, these variables could be important for understanding enlistment outcomes for gender as well as other demographic groups.

Included in the expression for the marginal productivity of effort (the $c_{sj}^{*}$ defined in Chapter Two for the three-contract-type model) were variables representing staffing, including numbers of production recruiters and their current status.[3] In the econometric analyses, num-

---

[2]   Since these data represented MOS distributions for the whole country, they can be viewed as exogenous from the perspective of individual stations. In future research, it would be preferable to utilize information based on available training slots.

[3]   These variables are the same as those used by Dertouzos and Garber (2006, Chapter Four); see that report for further details and discussion. Strictly speaking, the recruiter variables represent a measure of the command's effort or level of resources allocated for expand-

bers of OPRA recruiters were expressed by dichotomous indicator variables to allow for flexibility in the form of the relationship between number of recruiters and the marginal productivity of effort. During FYs 2001–2004, about 25 percent of all stations were staffed by a single production recruiter (the "left-out" or reference category or comparison benchmark in the regressions). Other variables were meant to capture staffing turbulence and the contributions of station commanders, expressed relative to numbers of OPRA recruiters. Included were percentages of assigned recruiters on duty but not on full production, number of recruiters assigned but temporarily on leave, and station commanders with limited or no responsibility to sign prospects.

Variables representing the presence and extent of reserve recruiting activity within the station were also included as potential determinants of the marginal productivity of effort directed at enlisting youth into the regular Army.[4] These variables are the number of reserve recruiters and the level of missions for prior service, high school, and other USAR enlistment categories. In principle, co-located USAR recruiting activity could increase or decrease the marginal productivity of efforts by OPRA recruiters to enlist youth for the regular Army. For example, positive spillovers from reserve recruiting effort to regular Army contract production could result from reserve recruiters' efforts to sell the military in general, or the Army in particular, as a career option. On the other hand, if reserve and active Army recruiting compete for a limited supply of eligible youth, a local reserve recruiting presence could cause the efforts of regular Army recruiters to be to less productive, other things being equal.

---

ing the supply of enlistees, holding an individual's level of effort constant. If, in fact, additional recruiters affected the level of individual effort through some other mechanism, such as free-riding (see the discussion in Chapter Five) then perhaps these variables should be included in the expression for individual recruiter effort as well. Since we do not know, a priori, the true functional form of either the effort or the market quality relationships, there exists no suitable test capable of distinguishing these alternative hypotheses. Suffice it to say that our efforts to include recruiter variables in the effort equation did not change the qualitative results overall and, more important, had virtually no effect on the PPM derivations and the resulting policy conclusions.

[4]  About 57 percent of Army recruiting stations are engaged in both regular Army and U.S. Army reserve (USAR) recruiting (Dertouzos and Garber, 2006, p. 35).

Several other dichotomous indicator variables were also included. These represented the month of the year to control for systematic seasonal variations in recruiting difficulty. Regional variables—indicating whether the recruiting station was located in the Mountain, North Central, South, or Pacific states—were also included to capture regional differences in factors such as propensity to enlist.

Information was also gathered on local climate, specifically, average temperature, total precipitation, and humidity index during July. For the sample, the July temperatures averaged 75 degrees and rainfall averaged 3.5 inches. The humidity index, which reflects the number of hours of "high humidity" in July, averaged over 57 with a standard deviation of 15.

Several market and demographic variables were also constructed and used to control for the marginal productivity of recruiter effort. These variables included the qualified military available population (QMA) per OPRA recruiter, expressed in natural logarithms. On average, per-recruiter QMA was about 650 youth. Also included were local economic factors, including the relative civilian-to-military wage rate and employment conditions (expressed as the logarithms of the unemployment rate and its month-to-month change). Demographic variables included population proportions of African Americans, Hispanics, men enrolled in college, residents of urban areas, and residents of population "clusters."[5] Also included were a measure of childhood poverty and a measure of growth in single-parent households. Finally, we also included measures of local populations reporting affiliation with a variety of organized religions, including Catholicism, Eastern religions (such as Buddhism and Hinduism), and Christian adherents other than Catholics.

---

[5] The U.S. Census Bureau defines *urban, rural* and *clustered* populations as follows (in answering the frequently asked question, "What is the difference between urban and rural population?"): ". . . urban areas . . . include urbanized areas and urban clusters. An urban area generally consists of a large central place and adjacent densely settled census blocks that together have a total population of at least 2,500 for urban clusters, or at least 50,000 for urbanized areas. . . . Rural . . . population [is population] not classified as urban." See U.S. Census Web site.

Several variables representing prevalence of all services' veterans in 2001 were also included to capture the effects of veterans as influencers and role models and local attitudes toward military service. Because these influences may be quite different depending on a veteran's era of service (World War II versus Vietnam versus the Persian Gulf War, for example), these measures were broken up into six subgroups based on veterans' ages in 2001. We suspect that, all things being equal, the presence of young veterans would be a positive influence or, at the very least, an indicator of a fertile recruiting environment. However, some cohorts of veterans (such as those who served in Vietnam) may not have as positive a view of their past military experience and might, for that reason, tend to discourage youth from enlisting.

Our final measure represents the strength of competition from the other services in the stations' local recruiting territories. Because the Army's current share in a given market (Army enlistments as a percentage of enlistments into all four services) is endogenous—i.e., it depends on the station's current contract production—the Army's share as of 1999 was included.[6] As can be seen from the last row of Table 3.1, the Army share averaged (across station areas and months during FYs 2001–2004) almost 35 percent of all enlistments with a standard deviation of almost 8, and (not reported in the table) the Army market share was between 27 and 43 for about 95 percent of the station months.

---

[6] Army share data were not available for about 10 percent of the station areas, and the missing data were replaced by predicted values derived from a regression equation. See Dertouzos and Garber (2006) for details. It is possible that the Army share captures persistent and unobserved market characteristics favoring different branches rather than competitive effects that are likely to depend on other Services' local resource expenditure. Regardless, this variable added to the explanatory power of the regression without affecting the estimates of other coefficients so we saw no reason to exclude it from the final regression.

## Estimates for the Graduate, Senior, and "Other" Contract Model

Nonlinear least-squares estimates of the parameters of the contract-production equations (2.9), and their standard errors (S.E.), are presented in Table 3.2. These regressions constrained the relative intercepts of the effort-per-recruiter equations (2.8) to equal the relative points associated with the three contract types used to reward recruiter production during our sample period. As discussed in Chapter Two, this is our preferred interpretation of recruiters' relative values of contracts of the three types. Specifically, these weights were 3, 2, and 1 for I-IIIA graduates, I-IIIA seniors, and other enlistments, respectively.[7]

We report more than 70 parameter estimates for each of the three contract types. In general, the results are consistent with findings from the high-quality model reported in Dertouzos and Garber (2006). For example, the findings indicate that higher recruiting goals (missions plus DEP losses) lead to higher levels of enlistment levels. In addition, past success substantially increases a station's responsiveness to mission changes. Market variables are quite important, including local economic conditions and demographics, seasonal and regional differences, and competition with the other services. However, there appear to be some interesting differences in these impacts across the three different contract types—differences that we discuss presently.

The first panel of estimates (Goal variables) reports the parameters of the effort-per-recruiter equations (2.8). The label "own" in the variable descriptions signifies that the variables that involve goals are computed

---

[7]  The Army's recruiter incentive program, which establishes eligibility for a series of awards (see Oken and Asch, 1997) and, ultimately, faster promotion (see Dertouzos and Garber, 2006, Chapter 6), provided 30 points for a high-quality graduate, 20 for a high-quality senior, and 10 for all other enlistments during this period. We consider alternative assumptions about the relative values of enlistment categories—and, implicitly, alternative estimates based on corresponding constrained values for the intercepts—in computing performance measures reported and analyzed in Chapter Four. The choice of these relative values, though crucial for the derivation and interpretation of PPMs, does not substantially affect the empirical significance of market factors in determining enlistments. Thus, we do not report the parameter estimates underlying the weights and construction of our less-preferred versions of the PPM (i.e., PPM1 and PPM3) considered in Chapter Four.

**Table 3.2**
**Estimated Determinants of Monthly, Station-Level Enlistments of High-Quality Graduates, High-Quality Seniors and Others, FYs 2001–2004**

| Independent Variable | Graduate Alpha Contracts | | Senior Alpha Contracts | | Other Contracts | |
|---|---|---|---|---|---|---|
| | Coefficient | S.E. | Coefficient | S.E. | Coefficient | S.E. |
| **Goal Variables** | | | | | | |
| Own goal (mission plus DEP loss) ($\beta_j$) | -0.0813 | 0.0142 | 0.2242 | 0.0064 | 0.0722 | 0.0111 |
| Own goal squared ($\gamma_j$) | 0.0112 | 0.0013 | 0.1231 | 0.0072 | 0.0106 | 0.0027 |
| Own goal x past success ($\beta_{jj}$) | 0.4673 | 0.0142 | 0.00001 | 0.00001 | 0.3282 | 0.0114 |
| Own goal squared x past success ($\gamma_{jj}$) | -0.0151 | 0.0015 | -0.00001 | 0.00001 | -0.0189 | 0.0035 |
| **Marginal-Productivity Variables (elements of $\alpha_j$)** | | | | | | |
| Constant | -0.1142 | 0.0296 | 0.1193 | 0.0014 | 0.3816 | 0.0933 |
| 2-Recruiter station | -0.0472 | 0.0033 | -0.0093 | 0.0001 | -0.0158 | 0.0095 |
| 3-Recruiter station | -0.0701 | 0.0038 | -0.0169 | 0.0002 | -0.0282 | 0.0105 |
| 4-Recruiter station | -0.0866 | 0.0041 | -0.0015 | 0.0000 | -0.0432 | 0.0116 |
| 5-Recruiter station | -0.0937 | 0.0045 | -0.0182 | 0.0002 | -0.0746 | 0.0126 |
| 6+ Recruiter station | -0.1024 | 0.0049 | -0.0083 | 0.0001 | -0.0805 | 0.0140 |
| Commander, on production | -0.0998 | 0.0036 | -0.0301 | 0.0003 | -0.1386 | 0.0111 |
| Recruiter on duty | 0.0978 | 0.0084 | 0.0022 | 0.0001 | 0.3420 | 0.0272 |
| Absent recruiter | 0.0715 | 0.0040 | -0.0266 | 0.0003 | 0.2891 | 0.0112 |
| Commanders, not on production | -0.0863 | 0.0079 | 0.0053 | 0.0001 | -0.1426 | 0.0270 |
| Reserve recruiters | 0.04569 | 0.00385 | -0.00023 | 0.00003 | 0.09209 | 0.01150 |
| Reserve mission, "other" | -0.00385 | 0.00361 | 0.01874 | 0.00022 | -0.02553 | 0.01060 |
| Reserve mission, prior service | -0.01733 | 0.00295 | 0.00726 | 0.00008 | 0.08474 | 0.00820 |
| Reserve mission, high school | -0.05764 | 0.00668 | -0.00041 | 0.00006 | -0.11207 | 0.02110 |

**Table 3.2—continued**

| Independent Variable | Graduate Alpha Contracts | | Senior Alpha Contracts | | Other Contracts | |
|---|---|---|---|---|---|---|
| | Coefficient | S.E. | Coefficient | S.E. | Coefficient | S.E. |
| DEP loss, "other" reserves | 0.02060 | 0.00655 | -0.01962 | 0.00023 | 0.05994 | 0.01800 |
| DEP loss, prior service reserves | 0.08601 | 0.04410 | 0.15125 | 0.04260 | -0.22330 | 0.12190 |
| DEP loss, high school reserves | 0.06069 | 0.00463 | 0.00159 | 0.00003 | 0.04960 | 0.01300 |
| February | 0.01030 | 0.0034 | 0.0046 | 0.0022 | 0.0154 | 0.0092 |
| March | 0.0171 | 0.0033 | 0.0128 | 0.0002 | 0.0337 | 0.0092 |
| April | 0.0154 | 0.0034 | -0.0204 | 0.0002 | 0.0618 | 0.0096 |
| May | -0.0013 | 0.0031 | -0.0686 | 0.0008 | -0.0057 | 0.0091 |
| June | 0.0305 | 0.0032 | -0.0753 | 0.0009 | 0.0037 | 0.0069 |
| July | 0.0218 | 0.0033 | -0.0178 | 0.0002 | -0.0142 | 0.0085 |
| August | 0.0249 | 0.0035 | -0.0198 | 0.0002 | 0.0026 | 0.0093 |
| September | 0.0151 | 0.0032 | -0.0531 | 0.0006 | 0.0420 | 0.0093 |
| October | 0.0279 | 0.0034 | 0.0071 | 0.0024 | 0.0645 | 0.0095 |
| November | -0.0057 | 0.0033 | 0.0141 | 0.0002 | 0.0258 | 0.0094 |
| December | -0.1142 | 0.0296 | 0.1193 | 0.0014 | 0.3816 | 0.0933 |
| Mountain | -0.0472 | 0.0033 | -0.0093 | 0.0001 | -0.0158 | 0.0095 |
| North Central | -0.0701 | 0.0038 | -0.0169 | 0.0002 | -0.0282 | 0.0105 |
| South | -0.0014 | 0.0032 | 0.0151 | 0.0024 | -0.0208 | 0.0091 |
| Pacific | 0.0088 | 0.0038 | -0.0024 | 0.0000 | 0.0085 | 0.0112 |
| Hot | 0.0019 | 0.0023 | -0.0043 | 0.0001 | -0.0542 | 0.0066 |
| Rain | 0.0364 | 0.0029 | -0.0004 | 0.0000 | 0.0254 | 0.0089 |
| Humidity | 0.0062 | 0.0031 | -0.0003 | 0.0001 | 0.0792 | 0.0101 |

Table 3.2 (Continued)

| Independent Variable | Graduate Alpha Contracts | | Senior Alpha Contracts | | Other Contracts | |
|---|---|---|---|---|---|---|
| | Coefficient | S.E. | Coefficient | S.E. | Coefficient | S.E. |
| QMA per recruiter | 0.04516 | 0.00158 | -0.00121 | 0.00003 | 0.08275 | 0.00496 |
| Unemployment change | 0.06072 | 0.00687 | 0.03278 | 0.00038 | 0.09438 | 0.02150 |
| Unemployment level | 0.01639 | 0.00220 | 0.01539 | 0.00018 | -0.01205 | 0.00666 |
| Relative wage | 0.02333 | 0.00514 | 0.01911 | 0.00022 | 0.15226 | 0.01530 |
| African American | -0.07242 | 0.00709 | -0.12388 | 0.00141 | 0.35636 | 0.02180 |
| Hispanic | 0.01258 | 0.00636 | 0.02306 | 0.00027 | 0.28711 | 0.01890 |
| College | 0.00083 | 0.00015 | 0.00114 | 0.00001 | -0.00078 | 0.00044 |
| Urban population | 0.02473 | 0.00491 | -0.01202 | 0.00015 | -0.05745 | 0.01660 |
| Cluster population | 0.02106 | 0.00779 | 0.05163 | 0.00060 | -0.09538 | 0.02740 |
| Growth in single parent homes | 0.01338 | 0.00234 | 0.01110 | 0.00013 | 0.01606 | 0.00724 |
| Poverty | -2.72052 | 0.18870 | -1.16747 | 0.01240 | 5.36934 | 0.67510 |
| Catholic | -0.03710 | 0.00718 | 0.02944 | 0.00034 | 0.00867 | 0.02140 |
| Eastern | -0.65630 | 0.04710 | -0.34254 | 0.00407 | -0.40633 | 0.33030 |
| Christian | -0.04334 | 0.00859 | -0.03507 | 0.00041 | 0.09025 | 0.02550 |
| Vet32 | 0.04539 | 0.02590 | -0.14274 | 0.00164 | 0.65262 | 0.07740 |
| Vet33-42 | 0.05818 | 0.01980 | 0.08425 | 0.00099 | -0.49109 | 0.06080 |
| Vet43-55 | 0.10068 | 0.01140 | 0.04968 | 0.00057 | 0.79049 | 0.03230 |
| Vet56-65 | -0.25122 | 0.01130 | -0.09389 | 0.00107 | -1.19608 | 0.03400 |
| Vet65-72 | 0.25291 | 0.01780 | 0.05316 | 0.00062 | 1.04609 | 0.05400 |
| Vet73 | -0.07062 | 0.00559 | 0.02582 | 0.00030 | -0.20117 | 0.01630 |
| Market share | 0.00070 | 0.00008 | 0.00018 | 0.00000 | 0.00128 | 0.00025 |

for the contract type corresponding to the dependent variable in that equation. For example, the reported coefficient of the graduate goal in the graduate equation (–0.0813) is an estimate of $\beta_G$. The estimated coefficient for the interaction between the goal and past performance in the graduate-alpha market (i.e., $\beta_{GG}$) is 0.4673. This estimate indicates that current-month station production for grad-alpha contracts is more responsive to mission increases for that contract type the more successful (relative to missions) the station has been in producing such contracts in the recent past.[8] In addition, the equations include quadratic terms in goal difficulty to allow for such nonlinear relationships as diminishing or even negative returns to increasing missions, as well as interactions with the past performance ratio. Thus, to ease interpretation, we have simulated the effects of increasing missions and increasing the variables assumed to determine marginal productivities.

Table 3.3 reports predicted effects on enlistments of a given type from increasing the station mission by one for three assumed levels of past success (the variables $R_G, R_S, R_O$ in the effort equations (2.8)), namely, their mean values and their mean values plus and minus one standard deviation. As can be seen from the table, graduate contracts are somewhat less sensitive to increases in their own missions than is the case for the other two enlistment categories. On average, an increase, of 1 in the high-quality graduate mission results in 0.29 additional contracts of that type. Senior-alpha contracts are on average about 17

---

[8]  The estimated effects of past performance could reflect to some degree econometric biases due to persistent but unobserved factors that are correlated with current production. (As is well-known, lagged dependent variables with serially correlated errors lead to bias in least-squares estimation.) The size of such critical correlations, if any, cannot be directly assessed quantitatively. However, it seems likely that a large portion of the variation in our measures of past success is due to randomness in production (see Chapter Five) as well as to the mission allocation process. To probe the sensitivity of our basic conclusions to this potential source of bias, we estimated versions of the model in which past performance was excluded and found that other coefficient estimates were not substantially affected. Most important, computed performance metrics based on this simplified model were highly correlated with those based on estimates of the more general model (which included the interaction between past performance and market difficulty).

**Table 3.3**
**Predicted Enlistment Increase Due to Increasing Missions for High-Quality Graduates, High-Quality Seniors, and Others**

|  | Graduates | Seniors | Others |
|---|---|---|---|
| Increase own mission by 1, past success is average | .29 | .34 | .38 |
| Increase own mission by 1, past success 1 SD above mean | .38 | .38 | .47 |
| Increase own mission by 1, past success 1 SD below mean | .21 | .30 | .28 |

percent more responsive to increases in own mission. Not surprisingly other contract categories are the most responsive to mission increases (because "others" are relatively easy to recruit). These results suggest that graduates are the most difficult category to recruit, at least on the margin, given the levels of grad-alpha missions during our sample period. This pattern could reflect various factors, such as lower average propensity[9] of graduates than for prospects of the other types or strong command-level preference for graduates (who typically spend less time in the delayed-entry pool), resulting in relatively ambitious missions for signing these prospects.

As can also be seen from Table 3.3, contracts are more responsive to mission increases of a given type the more successful the station has been in the recent past in enlisting prospects of that type. The differences in responsiveness due to differences in recent past success are considerable, suggesting that recent success is an important determinant of recruiter productivity. As we discuss in Dertouzos and Garber (2006, Chapter Four) drawing on literature in psychology and management, these effects may be due to several factors—greater success may increase recruiters' confidence in their abilities as recruiters, increase morale, or both. For example, an increase of one in the high-quality graduate mission in a previously successful market (i.e., a station territory for which our measure is one standard deviation above its mean) will result in a 0.38 increase in graduate I-IIIA contracts. This is about 80 percent higher than in markets that have experienced

---

[9]  In Army parlance, *propensity* refers to willingness to enlist, all other things being equal.

relatively low performance ratios for graduates (one standard deviation below the mean). This suggests that there are significant gains to be made by shifting graduate missions to markets that have experienced previous success in recruiting graduates. Similar results pertain to the senior-alpha and other enlistment categories, although the magnitude of the effects is somewhat smaller in the case of seniors.

Table 3.4 reports the predicted effects of different levels of the variables that influence the marginal productivity of recruiter effort in enlisting prospects of different types. To facilitate comparisons, enlist-

**Table 3.4**
**Impacts of Individual Market and Other Factors (Determinants of $c_j^*$) on High-Quality Graduate, High-Quality Senior, and Other Enlistments**

| Variable | Percentage Increase in Expected Enlistments Due to One-Standard-Deviation Increase in Variable | | |
|---|---|---|---|
| | Graduates | Seniors | Other |
| 2-recruiter station | −3.2 | −0.8 | −0.4 |
| 3-recruiter station | −4.4 | −1.9 | −0.6 |
| 4-recruiter station | −4.2 | 0.1 | −0.8 |
| 5-recruiter station | −3.0 | −0.7 | −0.9 |
| 6+ recruiter station | −1.9 | −0.1 | −0.6 |
| Reserve recruiters | 3.0 | −0.2 | 1.6 |
| Reserve mission, "other" | −0.5 | 1.7 | −0.4 |
| Reserve mission, prior service | −1.8 | 0.5 | 2.0 |
| Reserve mission, high school | −1.4 | 0.1 | −0.6 |
| DEP loss, "other" reserves | 0.6 | −0.7 | 0.5 |
| DEP loss, prior service reserves | 0.2 | 0.8 | −0.2 |
| DEP loss, high school reserves | 1.9 | 0.7 | 0.5 |
| February | 0.0 | −0.1 | 0.2 |
| March | 0.3 | 0.4 | 0.5 |
| April | 0.5 | −2.4 | 0.9 |
| May | −0.2 | −5.1 | −0.1 |
| June | 1.1 | −5.6 | 0.1 |

**Table 3.4—continued**

| Variable | Percentage Increase in Expected Enlistments Due to One-Standard-Deviation Increase in Variable | | |
| --- | --- | --- | --- |
| | Graduates | Seniors | Other |
| July | 0.7 | −1.8 | −0.2 |
| August | 0.8 | −1.9 | 0.0 |
| September | 0.4 | −4.5 | 0.6 |
| October | 1.2 | −3.8 | 1.1 |
| November | 0.1 | 0.8 | 0.4 |
| December | −0.2 | 0.7 | −0.3 |
| Mountain | 0.2 | 0.2 | 0.1 |
| North Central | 0.2 | −0.1 | −1.3 |
| South | 2.8 | 0.2 | 0.6 |
| Pacific | 0.1 | −0.2 | 1.5 |
| Hot | 1.4 | 0.5 | 0.4 |
| Rain | 0.3 | 1.5 | 1.8 |
| Humidity | −1.6 | −1.2 | −1.4 |
| Commander, on production | −5.1 | −1.7 | −2.5 |
| Recruiter on duty | 3.7 | 0.4 | 4.5 |
| Absent recruiter | 3.6 | −2.4 | 5.2 |
| Commanders, not on production | −3.2 | 0.3 | −2.0 |
| QMA per recruiter | 4.1 | −0.5 | 2.9 |
| Unemployment change | 0.7 | 0.4 | 0.6 |
| Unemployment level | 1.8 | 1.8 | −0.3 |
| Relative wage | 0.2 | 0.6 | 1.2 |
| African American | −2.0 | −4.7 | 2.8 |
| Hispanic | −0.3 | 0.5 | 2.9 |
| College | 0.6 | 1.6 | −0.2 |
| Urban population | 1.5 | −1.3 | −1.2 |
| Cluster population | 0.2 | 2.8 | −0.9 |
| Growth in single parent homes | 0.7 | 0.9 | 0.2 |
| Poverty | −1.9 | −1.2 | 1.2 |
| Catholic | −0.5 | 2.7 | 0.0 |
| Eastern | −0.5 | −0.1 | −0.2 |

**Table 3.4—continued**

| Variable | Percentage Increase in Expected Enlistments Due to One-Standard-Deviation Increase in Variable | | |
| --- | --- | --- | --- |
| | Graduates | Seniors | Other |
| Christian | −0.7 | 0.0 | 0.6 |
| Vet32 | 0.1 | −3.6 | 2.5 |
| Vet33-42 | 2.2 | 4.3 | −4.0 |
| Vet43-55 | 3.2 | 3.3 | 11.3 |
| Vet56-65 | −6.5 | −3.6 | −12.9 |
| Vet65-72 | 4.5 | 0.1 | 7.7 |
| Vet73 | −2.6 | 2.2 | −2.7 |
| Market share | 1.1 | 0.4 | 0.5 |

ment effects are expressed as percentage increases (from the respective sample means) due to a one-standard-deviation increase in the value of the explanatory factor. For graduates, the most important factors having a positive effect on contract production include QMA per OPRA recruiter, location in the south, and the size of the state's veteran populations between 33 and 55 years old and between 65 and 72. In contrast, graduate contracts are lower in poor neighborhoods or when there is an unusually high presence of Vietnam era veterans, aged 56 to 65. Staffing variables are also important. For example, the smallest stations (a single recruiter) perform better in signing high-quality graduates, all other things being equal.

These results are very similar to those reported in our earlier analysis of the market for high-quality enlistments (Dertouzos and Garber, 2006, Chapter Four). This should not be surprising given that the sample periods for the earlier and current analyses overlap considerably—the earlier study considered September 2001 through June 2003, and the current study analyzes data for FYs 2001–2004. In addition, high-quality graduates account for more than 70 percent of this market.[10] One notable exception is the estimated effect of the civilian/military wage ratio that, in the current study, did not have the expected

---

[10] For a more extensive discussion of the importance of individual factors, see Dertouzos and Garber (2006, pp. 76–89).

negative relationship that was found in our earlier study and many
other analyses of recruiter productivity. We determined that this surprising result is most likely due primarily to the time-series correlation
between military salary levels and unmeasured aspects of the quality
of the recruiting environment, particularly during the latter portion of
our sample period. As a result, we think that our estimated wage effects
are probably unreliable and are likely to have questionable validity in
previous studies as well.[11]

Observed patterns are somewhat different for the other two categories of contracts considered, I-IIIA seniors and other enlistees. For
example, a clustered population is neutral with respect to graduates
but favors high school seniors and is negatively correlated with all
other contract types. Catholic neighborhoods draw more high-quality
seniors but not graduates or others. Market areas with relatively large
proportions of African Americans and Hispanics are relatively good for
enlisting "other" youth, but not I-IIIA seniors and graduates. It is also
interesting that unemployment rates are more important to the supply
of high-quality youth than of other youth.

These results suggest that differences or variability in the marginal productivity of effort—across both station and contract types—
can substantially affect recruiters' opportunities to recruit youth from
different population segments. Such variability has important implications for methods of determining missions and for judging the productivity of stations.

----

[11] We explored this empirical anomaly by first reestimating our generalized model (equations (2.9)) using the data corresponding to the period analyzed in our earlier study (September 2001 through June 2003). The estimated wage effects were negative in these regressions.
Adding the data for the remainder of our current sample period (i.e., July 2003 through
September 2004), though not substantially affecting the other coefficients, changed the sign
of the military/civilian wage estimate. Running the model using station-level fixed effects
did not alter the results. However, using a monthly dummy variable—which eliminated the
time-series variation in the military wage—reversed the sign of the estimated wage effect
once again. We suspect that the anomalous sign of the wage effects reported in Table 3.4
stems from the resource allocation process in which military benefits are often increased
(decreased) in response to recruiting difficulties (success). As a result, studies relying on
time-series variations in benefit levels that do not consider this reverse causality may not be
reliable. This is an important issue for future research.

To reinforce this conclusion, Table 3.5 reports means, standard deviations, and correlations over station-months for our estimates of the marginal productivity of effort across the three contract types.[12] There is considerable variation in these measures. For example, the estimated marginal productivity of effort in enlisting high-quality graduates has a standard deviation of 0.242, which is more than 40 percent of the mean value. Clearly, a great deal of variation exists, from local market to local market, in the effort required to produce graduate contracts. The relative variation is even greater in the cases of seniors and lower-quality categories. Specifically, the standard deviations, expressed as percentages of the means, are over 60 and 70 percent for I-IIIA seniors and others, respectively.

More important, the correlation between any pair of these measures suggests that markets that are relatively good for recruiting prospects of one type may be relatively bad for recruiting prospects of other types. Indeed, the correlation between the marginal-productivity measures for I-IIIA graduates and seniors is virtually zero (0.023), and the correlation for seniors and others is negative. This suggests that characterizing recruiting markets by a single measure of market quality (or

**Table 3.5**
**Means, Standard Deviations and Correlations Between Estimated Marginal Productivities of Effort for High-Quality Graduates, High-Quality Seniors, and Others**

| Segment | Market Quality | Mean | S.D. | Correlation with Other Segments | | |
|---|---|---|---|---|---|---|
| | | | | $c_G^*$ | $c_S^*$ | $c_O^*$ |
| Graduates | $c_G^*$ | 0.552 | 0.242 | 1.000 | | |
| Seniors | $c_S^*$ | 0.140 | 0.087 | 0.023 | 1.000 | |
| Other | $c_O^*$ | 0.218 | 0.155 | 0.531 | −0.275 | 1.000 |

---

[12] As discussed in Chapter Two, the rescaling of effort and skill involves linear transformations of marginal productivity. Hence, rescaling does not affect correlation coefficients.

marginal productivity of effort) could obscure important differences across markets. Recognizing and exploiting such differences in allocating missions to stations could enhance recruiter productivity appreciably at little or no budgetary cost.

Table 3.6 reports estimates reinforcing the conclusion that allocations of missions of each type could be matched more closely to marginal productivity of effort. In particular, the table reports correlations of our estimates of the marginal productivity of effort with assigned missions per recruiter. For graduates, the correlation is only 0.35. For seniors, the correlation is even lower, at 0.12.

Our previous research (Dertouzos and Garber, 2006) demonstrated that failure to account adequately for variations in market quality in setting high-quality missions has had a modest but significant dampening effect on recruiter productivity. Similarly, a mismatch between performance measures and effort plus skill can cause recruiters to exert less effort than they otherwise would. The new results reported here indicate that the markets for graduates, seniors, and other enlistees are distinct, with exogenous factors such as local economic conditions, local population characteristics, and recruiting resources—all of which vary across markets—affecting these segments in different ways. To enhance recruiter productivity, missions or performance measures should more closely reflect this variation.

**Table 3.6**
**Correlations Between Market Quality and Missions per Recruiter**

| Market Quality | Graduate Mission | Senior Mission | Other Mission |
|---|---|---|---|
| $c_G^*$ | 0.350 | 0.025 | −0.031 |
| $c_S^*$ | −0.132 | 0.119 | −0.480 |
| $c_O^*$ | 0.228 | 0.132 | 0.652 |

## Estimates for a Model Distinguishing Men and Women

In the previous section, we presented estimates for a model distinguishing the three enlistment categories that are explicitly missioned. The econometric results just discussed suggest that important, policy-relevant distinctions exist among those segments of the youth population. In principle, other segments of the youth population may also exhibit important differences in the levels and determinants of the marginal productivity of effort. Moreover, it is likely that the Army values particular subgroups of enlistees more highly than other subgroups and that variations in the prevailing demographic or economic circumstances lead to significant differences across stations' market territories in the ability of recruiters to enlist different categories of youth. For example, there are potentially important differences between male and female recruits, as well as distinctions among educational categories (for example, those with and without some college) and, perhaps, market segments defined by race or ethnicity.

As we have discussed, such differences should be considered when designing effective (equitable and efficient) performance metrics. In this section, we conduct an exploratory analysis focused on gender differences. Clearly, male and female enlistees are valued differently because of practical as well as regulatory limitations on the roles of women in combat. In addition, it seems plausible that there is considerable variation in local conditions that affect the relative quality of the market for recruiting men versus women.

To analyze such possibilities, we adapt the three-contract-type model to consider four contract categories defined by gender and quality. In particular, we assume that monthly effort per recruiter is determined by:[13]

---

[13] In our previous specification distinguishing among seniors, graduates, and others, we included a quadratic term to allow for possible diminishing or even negative marginal effects of mission difficulty on effort. We also included an interaction term to account for the possible impact of past performance. In the senior-grad estimations reported earlier, exclusion of these variables did not affect the qualitative results in a significant manner. We have no reason to believe that they would be important in analyzing gender differences and, because of the addition of MOS interactions, we wished to reduce model complexity. However, we

$$e_{im} = 1 + (\beta_m + \beta_{Cm}C_i + \beta_{Wm}W_i + \beta_{Bm}B_i)d_{im}, \qquad (3.1)$$

where

$m$ indexes contract types in month $i$ (i.e., high-quality men, high-quality women, all other men, and all other women)

$e_{im}$ denotes effort per recruiter in a month directed at one of four contract types

$d_{im}$ denotes the difficulty of making the goal for contract type $m$, which is defined analogously to mission-difficulty measures considered in Chapter Two (e.g., equation (2.3)).[14]

Note, however, that since missions are not broken down by gender, the $d_{im}$ are computed using the aggregate goal per OPRA recruiter for all (i.e., male plus female) high-quality and all other enlistments, divided by the marginal productivity of effort for one of four specific market segments (high- and low-quality men and high- and low-quality women). Moreover, because the ability to sign a contract will depend on MOS eligibility, the difficulty variables for high-quality and other enlistments are interacted with measures of the command-level distributions of prevailing job allocations by broad categories (the MOS categories $C$, $W$, and $B$, discussed presently), which are likely to have different implications for men and women. The products of a difficulty measure and a job-availability measure can be viewed as the number of missions that are earmarked to fill broad MOS categories.

For example, $C$ is the national percentage of all enlistees signing up for combat-support jobs in a given month. The next category, $W$,

---

have generalized the expression to allow for cross-effects of missions (e.g., missions of high-quality categories can influence effort directed at low-quality recruits).

[14] In the empirical work reported in this chapter, we generalize equation (3.1), allowing for spillovers or cross-effects between mission categories. In particular, we add four parameters of the same form as above, interacted with the mission of the other enlistment category. In other words, when $m = h$ for high-quality enlistments and $m = l$ for low quality, the effort expression for high quality becomes: $e_{ih} = 1 + (\beta_h + \beta_{Ch}C_i + \beta_{Wh}W_i + \beta_{Bh}B_i)d_{ih} + (\delta_h + \delta_{Ch}C_i + \delta_{Wh}W_i + \delta_{Bh}B_i)\bar{d}_{ih}$ where $\bar{d}_{ih}$ is the goal for the low-quality category of enlistments divided by market quality for high-quality enlistments $= g_{il}/c_{ih}^*$.

represents white-collar MOSs—jobs that typically have clerical, service, or professional analogs in the private sector. It is likely that such MOSs would have more appeal to women than to men. Finally, $B$ represents blue-collar occupations. Analogs to these jobs also exist in the private sector but would not be considered office or professional occupations. Women may serve in all these categories except combat arms (the omitted category).

We estimated separate contract equations for each of the four enlistment categories, using the same list of determinants of marginal productivity of effort that we used in the model distinguishing graduates, seniors, and others. Estimated coefficients and standard errors are reported in Table 3.7. For the most part, the estimates are consistent with the relationships previously estimated when combining male and female categories but distinguishing seniors from graduates. However, important differences did emerge. To summarize the differences, Table 3.8 uses regression coefficients and the distribution of explanatory variables to compute the percentage change in enlistments, by category, that would result from a one-standard-deviation increase in each of the explanatory factors.

For example, one-recruiter stations in rural areas with relatively low prevailing wages and relatively high percentages of minority populations appear to be relatively good for recruiting women in general, especially women in the bottom half of the AFQT distribution. All things being equal, areas with a high African-American population can be expected to yield almost 10 percent more "other" women but more than 4 percent fewer high-quality men. While small recruiting stations appear to be less successful at signing high-quality graduates, they manage to attract more than their share of other women. There are significant regional patterns as well. Stations located in the North Central and South regions attract fewer women, but they appear to exceed national averages in enlisting high-quality men. The opposite holds for stations located in the Pacific region.

In Table 3.9, we compare estimates of station-level marginal productivity of effort for the four market segments. More specifically, the table—which is analogous to Table 3.5—computes average monthly

**Table 3.7**
**Model Estimates for Male and Female Contracts by Enlistment Category**

| | HQ Men | | HQ Women | | Other Men | | Other Women | |
|---|---|---|---|---|---|---|---|---|
| | Coefficient | S.E. | Coefficient | S.E. | Coefficient | S.E. | Coefficient | S.E. |
| **Mission variables** | | | | | | | | |
| High-quality mission plus DEP | -0.2141 | 0.0277 | -0.0031 | 0.0268 | 0.2578 | 0.0224 | 0.0248 | 0.0226 |
| Other mission plus DEP | 0.0166 | 0.0036 | -0.1746 | 0.0456 | 0.0132 | 0.0029 | -0.0600 | 0.0384 |
| Combat support x HQ mission | 1.2896 | 0.1355 | 0.1851 | 0.0752 | -0.1041 | 0.1098 | 0.0165 | 0.0633 |
| Combat support x other mission | -0.4748 | 0.2009 | 0.1649 | 0.1180 | 0.1460 | 0.1628 | -0.1236 | 0.0993 |
| White-collar MOS x HQ mission | 0.8738 | 0.0839 | 0.2725 | 0.0474 | -0.4099 | 0.0680 | -0.0111 | 0.0399 |
| White-collar MOS x other mission | 0.2392 | 0.1358 | 0.2169 | 0.0789 | -0.0801 | 0.1100 | 0.3201 | 0.0665 |
| Blue-collar MOS x HQ mission | -0.4087 | 0.1359 | -0.2684 | 0.0989 | -0.4971 | 0.1101 | -0.0041 | 0.0833 |
| Blue-collar MOS x other mission | 0.5995 | 0.1745 | 0.6756 | 0.1645 | 0.5176 | 0.1413 | 0.3856 | 0.1385 |
| **Marginal productivity variables** | | | | | | | | |
| Constant | 0.0202 | 0.1083 | 0.0425 | 0.0566 | 0.6047 | 0.0878 | 0.1166 | 0.0477 |
| 2-recruiter station | -0.2526 | 0.0148 | -0.0002 | 0.0076 | -0.0665 | 0.0120 | 0.0173 | 0.0064 |
| 3-recruiter station | -0.3349 | 0.0161 | 0.0068 | 0.0083 | -0.0847 | 0.0130 | 0.0237 | 0.0070 |
| 4-recruiter station | -0.3754 | 0.0173 | 0.0074 | 0.0090 | -0.1013 | 0.0140 | 0.0315 | 0.0075 |
| 5-recruiter station | -0.4040 | 0.0183 | 0.0030 | 0.0095 | -0.1306 | 0.0148 | 0.0233 | 0.0080 |
| 6+ recruiter station | -0.4130 | 0.0196 | 0.0053 | 0.0102 | -0.1407 | 0.0159 | 0.0210 | 0.0086 |

Table 3.7—continued

|  | HQ Men | | HQ Women | | Other Men | | Other Women | |
|---|---|---|---|---|---|---|---|---|
|  | Coefficient | S.E. | Coefficient | S.E. | Coefficient | S.E. | Coefficient | S.E. |
| Commander, on production | -0.2218 | 0.0146 | -0.0421 | 0.0077 | -0.0969 | 0.0119 | -0.0298 | 0.0064 |
| Recruiter on duty | 0.2696 | 0.0305 | 0.0874 | 0.0159 | 0.1964 | 0.0247 | 0.0813 | 0.0134 |
| Absent recruiter | 0.1724 | 0.0126 | 0.0869 | 0.0066 | 0.1643 | 0.0102 | 0.0711 | 0.0055 |
| Commanders, not on production | -0.2234 | 0.0305 | -0.0488 | 0.0159 | -0.1290 | 0.0247 | -0.0431 | 0.0134 |
| Reserve recruiters | 0.1214 | 0.0131 | 0.0275 | 0.0069 | 0.0636 | 0.0106 | 0.0171 | 0.0058 |
| Reserve mission, "other" | -0.0341 | 0.0120 | -0.0221 | 0.0063 | -0.0251 | 0.0097 | -0.0204 | 0.0053 |
| Reserve mission, prior service | -0.0569 | 0.0093 | -0.0103 | 0.0049 | 0.0350 | 0.0075 | 0.0143 | 0.0041 |
| Reserve mission, high school | -0.1635 | 0.0237 | -0.0746 | 0.0124 | -0.0772 | 0.0192 | -0.0387 | 0.0104 |
| DEP loss, "other" reserves | 0.0354 | 0.0202 | 0.0073 | 0.0106 | 0.0513 | 0.0164 | 0.0131 | 0.0089 |
| DEP loss, prior service reserves | 0.4280 | 0.1397 | 0.1282 | 0.0730 | -0.0009 | 0.1132 | -0.0012 | 0.0614 |
| DEP loss, high school reserves | 0.1126 | 0.0147 | 0.0437 | 0.0077 | 0.0359 | 0.0119 | 0.0145 | 0.0065 |
| February | -0.0157 | 0.0114 | -0.0117 | 0.0060 | 0.0220 | 0.0093 | -0.0078 | 0.0050 |
| March | -0.0116 | 0.0113 | -0.0153 | 0.0059 | 0.0455 | 0.0092 | -0.0101 | 0.0050 |
| April | -0.0452 | 0.0118 | -0.0249 | 0.0062 | 0.0685 | 0.0095 | 0.0017 | 0.0052 |

**Table 3.7—continued**

| | HQ Men | | HQ Women | | Other Men | | Other Women | |
|---|---|---|---|---|---|---|---|---|
| | Coefficient | S.E. | Coefficient | S.E. | Coefficient | S.E. | Coefficient | S.E. |
| May | -0.1322 | 0.0115 | -0.0441 | 0.0060 | 0.0258 | 0.0094 | -0.0085 | 0.0051 |
| June | -0.0074 | 0.0110 | -0.0173 | 0.0057 | -0.0201 | 0.0089 | -0.0060 | 0.0048 |
| July | -0.0202 | 0.0115 | -0.0192 | 0.0060 | -0.0124 | 0.0094 | -0.0058 | 0.0051 |
| August | 0.0038 | 0.0117 | -0.0046 | 0.0061 | 0.0150 | 0.0095 | -0.0089 | 0.0051 |
| September | 0.0721 | 0.0113 | 0.0071 | 0.0060 | 0.0439 | 0.0092 | 0.0126 | 0.0050 |
| October | 0.0233 | 0.0114 | 0.0037 | 0.0060 | 0.0490 | 0.0093 | 0.0036 | 0.0050 |
| November | -0.0421 | 0.0113 | 0.0051 | 0.0059 | 0.0068 | 0.0091 | 0.0007 | 0.0050 |
| December | -0.0329 | 0.0117 | -0.0027 | 0.0061 | -0.0244 | 0.0095 | -0.0064 | 0.0051 |
| Mountain | 0.0087 | 0.0137 | -0.0084 | 0.0072 | 0.0488 | 0.0111 | -0.0042 | 0.0060 |
| North Central | 0.0831 | 0.0038 | -0.0189 | 0.0046 | -0.0544 | 0.0071 | -0.0356 | 0.0039 |
| South | 0.1111 | 0.0111 | 0.0079 | 0.0058 | 0.0087 | 0.0090 | -0.0218 | 0.0049 |
| Pacific | -0.0330 | 0.0119 | 0.0301 | 0.0062 | 0.0690 | 0.0096 | 0.0311 | 0.0052 |
| Hot | 0.00037 | 0.00003 | 0.00004 | 0.00002 | 0.00000 | 0.00003 | -0.00002 | 0.00001 |
| Rain | -0.00006 | 0.00002 | 0.00007 | 0.00001 | 0.00014 | 0.00002 | 0.00010 | 0.00001 |
| Humidity | -0.00160 | 0.00023 | -0.00066 | 0.00012 | -0.00151 | 0.00019 | -0.00034 | 0.00010 |
| QMA per recruiter | 0.0937 | 0.0058 | 0.0090 | 0.0030 | 0.0578 | 0.0047 | -0.0013 | 0.0025 |
| Unemployment change | 0.0565 | 0.0255 | 0.0226 | 0.0133 | 0.0361 | 0.0207 | 0.0016 | 0.0112 |
| Unemployment level | 0.0522 | 0.0077 | 0.0052 | 0.0040 | 0.0328 | 0.0063 | 0.0034 | 0.0034 |

**Table 3.7—continued**

| | HQ Men | | HQ Women | | Other Men | | Other Women | |
|---|---|---|---|---|---|---|---|---|
| | Coefficient | S.E. | Coefficient | S.E. | Coefficient | S.E. | Coefficient | S.E. |
| Relative wage | 0.0131 | 0.0179 | 0.0172 | 0.0094 | 0.1460 | 0.0145 | 0.0324 | 0.0079 |
| African American | -0.6133 | 0.0249 | 0.0285 | 0.0130 | 0.1262 | 0.0202 | 0.2643 | 0.0110 |
| Hispanic | -0.1177 | 0.0218 | 0.0616 | 0.0114 | 0.1483 | 0.0177 | 0.1291 | 0.0096 |
| College | 0.0011 | 0.0005 | 0.0005 | 0.0003 | -0.0005 | 0.0004 | 0.0002 | 0.0002 |
| Urban population | 0.0965 | 0.0196 | -0.0516 | 0.0103 | -0.0176 | 0.0159 | -0.0614 | 0.0086 |
| Cluster population | 0.0691 | 0.0326 | 0.0040 | 0.0170 | -0.0914 | 0.0264 | -0.0373 | 0.0143 |
| Growth in single parent homes | 0.0569 | 0.0080 | 0.0187 | 0.0042 | -0.0018 | 0.0065 | 0.0067 | 0.0035 |
| Poverty | -7.6123 | 0.7632 | -2.9135 | 0.3986 | 6.0582 | 0.6183 | -0.2669 | 0.3356 |
| Catholic | -0.1799 | 0.0277 | 0.0212 | 0.0145 | -0.0722 | 0.0225 | 0.0282 | 0.0122 |
| Eastern | -0.5595 | 0.3803 | 0.3391 | 0.1986 | 0.0122 | 0.3081 | -0.4835 | 0.1672 |
| Christian | -0.1858 | 0.0291 | -0.0631 | 0.0152 | 0.0962 | 0.0236 | -0.0223 | 0.0128 |
| Vet32 | -0.0060 | 0.0920 | 0.1949 | 0.0480 | -0.0789 | 0.0745 | 0.2476 | 0.0404 |
| Vet33-42 | 0.1589 | 0.0712 | 0.0351 | 0.0372 | -0.1800 | 0.0577 | -0.1483 | 0.0313 |
| Vet43-55 | 0.1165 | 0.0391 | 0.2101 | 0.0204 | 0.4209 | 0.0317 | 0.2239 | 0.0172 |
| Vet56-65 | -0.3677 | 0.0435 | -0.3922 | 0.0227 | -0.6699 | 0.0352 | -0.3648 | 0.0191 |
| Vet65-72 | 0.3199 | 0.0655 | 0.2543 | 0.0342 | 0.7485 | 0.0531 | 0.3064 | 0.0288 |
| Vet73 | -0.1068 | 0.0202 | -0.0342 | 0.0106 | -0.1835 | 0.0164 | -0.0463 | 0.0089 |
| Market share | 0.0018 | 0.0003 | 0.0006 | 0.0001 | 0.0005 | 0.0002 | 0.0006 | 0.0001 |

**Table 3.8**
**Estimated Effects of Market and Other Factors on Male and Female Enlistments**

| Factor | Change Due to One Standard Deviation of Factor (%) | | | |
|---|---|---|---|---|
| | HQ Men | HQ Women | Other Men | Other Women |
| 2-recruiter station | −5.7 | 0.0 | −2.4 | 2.0 |
| 3-recruiter station | −6.7 | 0.5 | −2.7 | 2.5 |
| 4-recruiter station | −6.0 | 0.4 | −2.5 | 2.6 |
| 5-recruiter station | −4.2 | 0.1 | −2.1 | 1.3 |
| 6+ recruiter station | −2.5 | 0.1 | −1.3 | 0.7 |
| Reserve recruiters | 1.8 | 1.6 | 1.5 | 1.4 |
| Reserve mission, "other" | −0.5 | −1.2 | −0.5 | −1.5 |
| Reserve mission, prior service | −1.1 | −0.8 | 1.1 | 1.6 |
| Reserve mission, high school | −0.8 | −1.4 | −0.6 | −1.0 |
| DEP loss, "other" reserves | 0.3 | 0.2 | 0.5 | 0.5 |
| DEP loss, prior service reserves | 0.4 | 0.4 | 0.0 | 0.0 |
| DEP loss, high school reserves | 1.1 | 1.6 | 0.5 | 0.7 |
| February | −0.2 | −0.6 | 0.4 | −0.5 |
| March | −0.2 | −0.8 | 0.8 | −0.7 |
| April | −0.6 | −1.2 | 1.2 | 0.1 |
| May | −1.7 | −2.2 | 0.4 | −0.6 |
| June | −0.1 | −0.9 | −0.4 | −0.4 |
| July | −0.3 | −1.0 | −0.3 | −0.4 |
| August | 0.0 | −0.2 | 0.2 | −0.6 |
| September | 0.9 | 0.4 | 0.8 | 0.9 |
| October | 0.3 | 0.2 | 0.9 | 0.2 |
| November | −0.5 | 0.3 | 0.1 | 0.0 |
| December | −0.4 | −0.1 | −0.4 | −0.4 |
| Mountain | 0.1 | −0.4 | 0.9 | −0.3 |
| North Central | 1.6 | −1.4 | −1.6 | −3.8 |
| South | 2.5 | 0.7 | 0.3 | −2.6 |

**Table 3.8—continued**

| Factor | Change Due to One Standard Deviation of Factor (%) | | | |
| --- | --- | --- | --- | --- |
| | HQ Men | HQ Women | Other Men | Other Women |
| Pacific | −0.5 | 1.9 | 1.6 | 2.7 |
| Hot | 1.3 | 0.6 | 0.0 | −0.4 |
| Rain | −0.6 | 2.4 | 1.9 | 4.8 |
| Humidity | −1.1 | −1.8 | −1.5 | −1.3 |
| Commander, on production | −3.4 | −2.5 | −2.2 | −2.4 |
| Recruiter on duty | 3.1 | 3.9 | 3.4 | 5.0 |
| Absent recruiter | 2.7 | 5.2 | 3.8 | 5.9 |
| Commanders, not on production | −2.7 | −2.2 | −2.3 | −2.7 |
| QMA per recruiter | 2.7 | 1.0 | 2.5 | −0.2 |
| Unemployment change | 0.3 | 0.5 | 0.3 | 0.0 |
| Unemployment level | 0.9 | 0.3 | 0.8 | 0.3 |
| Relative wage | 0.1 | 0.5 | 1.5 | 1.2 |
| African American | −4.2 | 0.7 | 1.3 | 9.7 |
| Hispanic | −1.1 | 2.1 | 1.9 | 6.1 |
| College | 0.2 | 0.4 | −0.2 | 0.3 |
| Urban population | 1.7 | −3.6 | −0.4 | −5.9 |
| Cluster population | 0.6 | 0.1 | −1.1 | −1.7 |
| Growth in single-parent homes | 0.7 | 0.9 | 0.0 | 0.4 |
| Poverty | −1.5 | −2.2 | 1.8 | −0.3 |
| Catholic | −1.2 | 0.5 | −0.7 | 1.0 |
| Eastern | −0.2 | 0.4 | 0.0 | −0.8 |
| Christian | −1.1 | −1.5 | 0.9 | −0.7 |
| Vet32 | 0.0 | 2.5 | −0.4 | 4.5 |
| Vet33-42 | 1.2 | 1.0 | −1.8 | −5.6 |
| Vet43-55 | 1.3 | 10.0 | 7.6 | 14.8 |
| Vet56-65 | −3.3 | −14.0 | −9.1 | −18.1 |
| Vet65-72 | 2.0 | 6.3 | 7.0 | 10.5 |
| Vet73 | −1.2 | −1.6 | −3.2 | −2.9 |
| Market share | 0.7 | 0.8 | 0.3 | 1.1 |

Table 3.9
**Correlation Between Market Quality for Male and Female Enlistments**

| Quality of Market | Mean | S.D. | Correlations with Other Segments | | | |
|---|---|---|---|---|---|---|
| | | | $C*_{mh}$ | $C*_{fh}$ | $C*_{ml}$ | $C*_{fl}$ |
| High-quality men, $C*_{mh}$ | 0.562 | 0.231 | 1.000 | | | |
| High-quality women, $C*_{fh}$ | 0.163 | 0.077 | 0.745 | 1.000 | | |
| Other men, $C*_{ml}$ | 0.350 | 0.145 | 0.670 | 0.750 | 1.000 | |
| Other women, $C*_{fl}$ | 0.064 | 0.066 | 0.006 | 0.521 | 0.554 | 1.000 |

estimates of the marginal productivity of effort for all stations using the coefficient estimates reported in Table 3.7. As is apparent from the reported standard deviations, there is considerable variation in estimated levels of marginal productivity. Judging from the standard deviations relative to the corresponding means (i.e., coefficients of variation), there appears to be much more variation for other women than for the other three categories of enlistees. Also of interest are the somewhat weak correlations between some pairs of the market quality measures. For example, while the correlation for high-quality men and women is 0.745, there is virtually no correlation between high-quality men and other women.

The estimates also indicate that missioning and MOS availability can have important effects on both the volume and the distribution of enlistments. In Table 3.10, we present the results of simulations based on coefficient estimates that were presented in Table 3.7 and the mean values of the variables contributing to the marginal productivity of effort. The first row of the table contains predictions of the average impact of increasing the high-quality mission (senior and grad alphas) by one, holding the distribution of available MOSs constant. In response, recruiters increase effort directed at high-quality enlistments and, as a result, sign about 0.25 more high-quality contracts, with about 75 percent of this increase involving men. This percentage is slightly lower than the overall percentage of high-quality men among total high-quality contracts (89 percent), suggesting that it is somewhat easier to recruit additional high-quality women than men.

**Table 3.10**
**Impact of Missioning on Male and Female Enlistments**

| | Increase in Contracts | | | | |
|---|---|---|---|---|---|
| | HQ Men | HQ Women | Other Men | Other Women | Total |
| Increase HQ mission by 1 | 0.19 | 0.06 | 0.09 | 0.03 | 0.36 |
| Increase other mission by 1 | 0.10 | 0.04 | 0.23 | 0.08 | 0.45 |

In addition, increases in the "other" categories also occur. This may be due to recruiters substituting easier to enlist "others" when unable to meet the higher high-quality target. However, it may simply be the case that recruiters are unable to identify higher-qualified candidates with precision, especially at the earlier stages of the recruiting process.

Table 3.10 also reports that, in response to an increase by one in the other mission, contracts signed by other men and other women would be predicted to increase by 0.23 and 0.08, respectively. Higher-quality male and female enlistees will also increase, but by smaller amounts. In sum, an increase of one in the other mission, on average, is predicted to result in a 0.45 increase in total contracts, about 27 percent of which (i.e., 12/45) will be signed by women. This percentage is about 6 percentage points higher than the share of women in total enlistments. As in the case of high-quality-mission increases, mission increases are more likely to lead to more enlistments of women, probably because they are easier to recruit at the margin.

Changes in the distribution of job types, holding total mission constant, could have an effect on the number and composition of enlistments. For example (see Table 3.11), combat support and white-collar MOSs attract more high-quality men and women relative to combat jobs. Indeed, these effects are quite large. A 2-percent expansion in these desirable occupations can increase enlistments by an equivalent number.[15] Blue-collar MOSs are preferable to combat, but the market

---

[15] The estimated expansion effects by sector seem large but are plausible given the specification of the econometric model. For example, if high-quality women represent about 15 percent of the enlistment force, a 6.8 percent expansion in this category represents only 1 percent of the total force. Also, recall that the MOS distributions can be interpreted as the percentage of the mission that is allocated to different occupations. Historically, only about

**Table 3.11**
**Impact of MOS Distribution on Male and Female Enlistments**

| | Increase in Contracts Relative to Mean (%) | | | | |
|---|---|---|---|---|---|
| | HQ Men | HQ Women | Other Men | Other Women | Total |
| Increase combat support 2% | 4.46 | 4.78 | −0.12 | −0.99 | 2.31 |
| Increase white-collar MOS 2% | 4.38 | 6.79 | −2.38 | 3.17 | 2.10 |
| Increase blue-collar MOS 2% | −0.32 | 1.87 | 0.08 | 3.99 | 0.49 |

expansion effect is much smaller. Such jobs appeal to different market segments in quite different ways. Surprisingly, blue-collar occupations appeal to other women, perhaps because there are private-sector barriers to these careers. A larger proportion of white-collar jobs are positively related to numbers of enlistments by both categories of women and by high-quality men. However, these MOSs are not as attractive to other men.

Finally, in Table 3.12, we present simulations assuming increasing numbers of recruiters. The first row indicates the predicted average gross contracts by category, computed at the mean values for the data set. We then simulate the marginal impact of adding 10 percent more recruiters and distributing these recruiters uniformly across all size categories of recruiting stations.[16] The simulation results reported in the first row of the table hold mission constant and increase the number of recruiters only. As a result, the mission is rendered less difficult, and per-recruiter effort is predicted to fall in response to the increase in recruiters. Still, enlistments in all categories are predicted to increase, although the simulations clearly indicate diminishing returns. This is because the estimated elasticity of contracts in each category is approximately 0.5. In other words, contracts increase only by about 5

---

80 percent of the mission is achieved. Thus, a 2.3 percent increase in contracts is slightly smaller, in absolute numbers, than a 2 percent increase in available white-collar mission slots (.8 × 2.3 = 1.8 percent).

[16] The 10-percent increase in recruiters was distributed so that all station size categories experienced the same average percentage increase in recruiters. Specifically, one in ten of the one-recruiter stations received an additional recruiter, two of ten two-recruiter stations received an additional recruiter, and so forth.

Table 3.12
**Male and Female Enlistment Increases Due to Added Recruiters**

|  | HQ Men | HQ Women | Other Men | Other Women |
|---|---|---|---|---|
| Average contracts per recruiter | 0.932 | 0.335 | 0.605 | 0.156 |
| Marginal percentage increase (%) |  |  |  |  |
| Add 10% recruiters, allow effort to vary | 5.16 | 4.87 | 5.50 | 5.16 |
| Add 10% recruiters, hold effort constant in each category | 7.97 | 9.25 | 10.22 | 11.11 |

percent when recruiters are increased by 10 percent. These results suggest that there are diminishing returns to adding recruiters, and these declines are similar across AFQT categories and gender.

In contrast, the last row of simulation results in Table 3.12 assumes that missions increase along with the number of recruiters so that mission difficulty is held constant despite the increased potential supply stimulated by the extra recruiters. This is achieved by increasing missions to balance the increase in productivity induced by the larger staff. Since mission difficulty remains constant, the model predicts that effort in each category (and, therefore, total effort) will be unchanged. As a result, the marginal increases in enlistments will be greater than when effort is allowed to adjust. Note that for high-quality men, the elasticity increases but remains less than one. In other words, there are diminishing returns to adding recruiters, at least with respect to high-quality men. However, by adjusting mission and inducing constant effort, productivity is almost 60 percent higher. Note that the elasticities are much higher for women of both quality categories as well as for lower-quality men. In fact, when missions are set appropriately, there is no evidence of diminishing returns to recruiters.

These recruiter effects can be translated into a variety of marginal cost estimates. Under the assumption that a recruiter costs approximately $3,000 a month, the addition of recruiters can generate high-quality enlistments that cost about $5,300 on the margin.[17] This

---

[17] A 10-percent increase in the number of recruiters (.10 × 2.48 = .248) will increase high-quality enlistments by .14 (men plus women). At a monthly cost of $3,000, the marginal cost is computed as ($3,000 × .248)/.14 or $5,300 per additional recruit.

assumes that effort is allowed to fall because missions are not simul-
taneously increased. In contrast, if missions are increased to induce
greater effort, the marginal cost is less than $3,100.

Of course, the Army benefits from an enhanced supply of lower-
quality contracts as well. Expressed in terms of total manpower, an
additional contract costs only $3,300, even without increasing mis-
sions. With a concurrent mission increase, the cost per additional contract falls to just under $1,775. Although our model does not enable
a precise calculation of the ability to trade off effort among categories of recruits, it is clear that the expansion of less-desirable contracts
would enable recruiters to focus more effort on high-quality recruits.
The assumption that enlisting lower-quality youth takes 25 percent less
effort than enlisting I–IIIA seniors and graduates implies that the marginal cost of high-quality enlistees could be as low as $4,200, allowing
total effort to fall, or $2,500 if effort is constrained to stay the same.[18]

These results have several implications for performance measurement. First, markets differ significantly in the difficulty of enlisting
various segments of the youth population. As we have seen, gender differences are pronounced, and we suspect that other subsets of enlistees
are equally important to distinguish for performance evaluation. Not
only do such differences reflect systematic and fixed market differences
among stations, they are also likely to change significantly over time
as market conditions, the allocation of recruiting resources, the level
of missions, and the MOS composition required of accessions change.
Ideally, a performance measure would consider such market differences
if comparisons over time and among organizational units are to remain
meaningful.

---

[18] We have also seen that a mission increase for lower-quality categories generally results in
25 percent more enlistments than the increase in high-quality enlistments due to increasing the high-quality mission. In addition, holding effort constant, an increase in recruiters
results in 25 percent more enlistments in lower-quality categories. So, as an approximation,
we can assume that the relative difficulty of high-quality categories is at least 25 percent.

# CHAPTER FOUR

# Empirical Analysis of Performance Measures

In this chapter, we consider performance measurement empirically. We do so by focusing on station-level performance measurement for FY 2004 and the three categories of enlistment contracts that were missioned during our analysis period, namely, high-quality (AFQT I-IIIA) graduates, high-quality seniors, and all others. More specifically, we compute and analyze five traditional measures of recruiting performance for FY 2004 and compare them with three versions of our conceptually preferred performance metric that was presented in equation (2.10). All eight of these measures are based on numbers of contracts signed, but they differ in other important respects.

The five traditional measures, which in principle can be computed for time periods (or *performance windows*) of various lengths such as months or quarters, are the following:

- Total contracts (over the three contract types) per OPRA recruiter, which is often called the *total write rate* (TWR)
- High-quality contracts (the sum of AFQT I-IIIA high-school graduates and high-school seniors enlisted) per OPRA recruiter, which is often called the *high-quality write rate* (HWR)
- Total contracts (over the three contract types) divided by the corresponding *total mission* (TM)
- High-quality contracts divided by *high-quality mission* (HM)
- The number of months during FY 2004 the station achieved the *regular Army mission box* (BOX).

These traditional measures can be computed from readily available data on contracts, missions, and numbers of OPRA recruiters.

In contrast, computation of our preferred performance metrics—which estimate the effort plus skill applied per OPRA recruiter to produce the contracts actually written—requires estimates from econometric analyses of contract production. To recapitulate, the motivation for considering these less easily computed measures (i.e., our PPMs) is that traditional performance measures are sensitive to factors that recruiters cannot control—most importantly, the difficulty of recruiting youth of different types in their stations' market territories. As a result, the traditional measures are not equitable and also fail to provide appropriate incentives for recruiters to put forth effort and for soldiers who are well suited to recruiting duty to volunteer for recruiting. Reducing this sensitivity to variations in recruiting difficulty requires measures or estimates of difficulty for each station for each contract type. Finally, since recruiting difficulty by station area cannot be measured directly, it must be estimated from observations on contract levels, numbers of recruiters, characteristics of market areas, etc. In sum, no readily computed measure—either the five we consider empirically or any others—can be considered conceptually satisfying for providing incentives for exerting effort or volunteering for recruiting duty, and none can be considered equitable.

While all traditional measures are conceptually flawed, the degrees to which they are flawed vary across the five traditional measures that we analyze empirically. More specifically, write rates (TWR and HWR) are measures of production per recruiter that make no direct, or even implicit, adjustments for market quality. In contrast, measures of contracts relative to missions (TM, HM and BOX) do adjust implicitly, but incompletely,[1] for variations in market quality to the extent that allocations of missions across stations accurately reflect variations in local conditions. Finally, BOX suffers from a flaw that does not pertain to TM or HM, namely, discontinuities due to the

---

[1] As was demonstrated in Dertouzos and Garber (2006), missions do partially adjust for market quality, but those adjustments are less than complete.

award system that provides substantial numbers of bonus points for making mission.

The three versions of our PPM involve alternative weighting schemes for combining, for each station, contracts of the three separately missioned types. Conceptually, the weights—see equation (2.10) and its discussion—are station-specific levels of the difficulty of recruiting youth in the three contract categories. Empirically, the weights—which are estimates of relative difficulty or, alternatively, the expected level of effort plus skill required to produce an enlistment—differ across the three versions of the PPM. The weights differ because they use alternative econometric estimates, which in turn are based on different assumptions about the relative values (in terms of utility) that recruiters place on producing the three different types of contracts. We interpret these recruiter preferences as resulting from the guidance of their leaders, which in turn reflects the relative values that the recruiting command places on the different types of contracts.

Our first PPM, PPM1, is based on assuming the same implicit value (or preference) for each category of enlistees. The second, PPM2, is based on assuming that the relative value or preference for categories of enlistees equals the relative points assigned by the recruiter award system over this period: three points for I-IIIA graduates, two points for I-IIIA seniors, and one point for all other categories of enlistments. This version and the derived measure are based on the estimates reported in Table 3.2. Finally, PPM3 values high-quality enlistees only and equally.[2] In our view, these alternative assumptions span the entire plausible range. For example, it is not plausible that low-quality contracts would ever be more highly valued than high-quality ones. PPM2, which can be viewed as a compromise between PPM1 and PPM3, is our most-preferred metric. This is because award points seem to us to be the most plausible and defensible expression of the preferences of the command as communicated to and internalized by recruiters in the field. We consider all three PPMs empirically, however, to enable examination of the

---

[2]   To conserve space, we do not report the estimates used in constructing PPM1 and PPM3 (i.e., the analogs to the estimates reported in Table 3.2).

sensitivity of stations' performance rankings to alternative, plausible assumptions.

Our empirical analysis focuses on the question: How well can traditional performance metrics—or even combinations of them— approximate PPM2, our most-preferred measure of recruiting station performance during FY 2004? The answer to this question is crucial because, for example, if a traditional measure or a combination of them can closely approximate PPM2, then the Army can measure performance satisfactorily without having to use—and periodically update— econometric estimates.

We first consider this question using the rank correlation coefficients for all pairs of the eight performance measures reported in Table 4.1. We utilized data from 1,417 stations with complete data for FY 2004. For these comparisons, we used rank (rather than ordinary) correlation coefficients for two reasons. First, station rankings are of considerable direct interest. For example, Army recruiting leaders may be interested in such questions as whether a particular station is in the top 10 percent in terms of performance or whether station X is more productive than station Y. Second, rank correlations are less sensitive than ordinary correlations to outlier values of estimated PPMs caused by

**Table 4.1**
**Rank Correlation Coefficients Among Five Traditional and Three Preferred Station-Level Performance Measures**

|       | TWR   | HWR   | TM    | HM    | BOX   | PPM1  | PPM2  | PPM3 |
|-------|-------|-------|-------|-------|-------|-------|-------|------|
| TWR   | 1     |       |       |       |       |       |       |      |
| HWR   | 0.825 | 1     |       |       |       |       |       |      |
| TM    | 0.774 | 0.685 | 1     |       |       |       |       |      |
| HM    | 0.654 | 0.816 | 0.859 | 1     |       |       |       |      |
| BOX   | 0.545 | 0.646 | 0.757 | 0.822 | 1     |       |       |      |
| PPM1  | 0.568 | 0.545 | 0.528 | 0.515 | 0.421 | 1     |       |      |
| PPM2  | 0.577 | 0.634 | 0.555 | 0.608 | 0.487 | 0.936 | 1     |      |
| PPM3  | 0.488 | 0.653 | 0.534 | 0.676 | 0.545 | 0.686 | 0.800 | 1    |

NOTE: Includes 1,417 stations with complete data for FY 2004.

implausible estimates of the difficulty of recruiting youths of specific types in particular station areas.[3]

Note first from Table 4.1 that the traditional measures are all positively—and moderately—correlated with one another. More specifically, all the rank correlations are at least 0.545 (in the case of TWR and BOX), with the next lowest being 0.646 (for HWR and BOX). The highest correlation between any two of our traditional measures is 0.859 (for TM and HM). Most important, the rank correlations between the traditional measures and PPM2 range from 0.487 (for BOX) to 0.634 (for HWR). In our view, none of these correlations is high enough to indicate that a traditional measure can provide a good approximation to PPM2. We then considered whether the PPM could be well approximated by a linear combination of the five traditional measures by computing a least-squares regression of PPM2 on a constant, TWR, HWR, TM, HM, and BOX.[4] The $R^2$ statistic from this regression was 0.659, which implies a correlation that is somewhat higher than the simple correlation coefficients between PPM and each of the traditional measures.[5] Finally, note that the correlations between the five traditional measures and our alternative preferred measures (i.e., PPM1 and PPM3) are in the same range as the corresponding correlations involving PPM2. In sum, it appears that traditional performance measures fall considerably short of being equitable and providing good incentives for recruiter effort and for soldiers with good sales skills or aptitudes to volunteer for recruiting.

It is worth noting that the alternative PPMs are not highly correlated in each comparison. Not surprisingly, this implies that the ranking of performance depends on how the separate categories are valued by USAREC. However, the correlation between PPM1 and PPM2,

---

[3] For example, some of the station-specific estimates of the marginal productivity of effort (i.e., some of the estimated $c_{sj}^{*}$) are negative or implausibly small (presumably due to sampling error).

[4] As is well known, an ordinary least squares regression chooses regression coefficients so that combining the independent variables using these coefficients provides the linear combination that has the highest correlation with the dependent variable.

[5] Note that the $R^2$ statistic represents the square of the correlation coefficient, so that an $R^2$ of 0.659 implies that R= $\sqrt{.659}$ = 0.812.

both of which place value on both high- and low-quality enlistees, is quite high, at 0.936.

The fairly low correlations between traditional measures and the three PPMs that we have just discussed suggest that performance rankings based on traditional metrics are likely to be very misleading. To illustrate how misleading the use of traditional measures can be, we compared station rankings, by quartiles (i.e., quarters of their distributions) for the frequency of making regular Army mission box during FY 2004 (i.e., BOX) to rankings based on PPM2. The comparison of station measured-performance quartiles based on BOX and PPM2 is summarized in Table 4.2.[6]

For example, the first row of Table 4.2 pertains to stations that ranked in the first quartile (top 25 percent) based on making mission box during 2004 (i.e., BOX) and reports the percentage of those stations that fall into the various quartiles of the PPM2 distribution. If BOX were an accurate guide to rankings based on PPM2, most stations falling into the top quartile of BOX would also be in the highest quartile for PPM2. Note, however, that only 46.3 percent of the stations ranked in the top quarter by BOX were also in the top quarter

**Table 4.2**
**Quartile Rankings Based on Preferred Performance Metric Versus Making Mission Box**

| Making Mission (BOX) | Preferred Performance Metric (PPM2), % | | | |
|---|---|---|---|---|
| | Top 25% | 2nd 25% | 3rd 25% | Bottom 25% |
| Top 25% | 46.3 | 35.0 | 12.7 | 5.9 |
| 2nd 25% | 29.7 | 27.4 | 29.9 | 13.3 |
| 3rd 25% | 16.1 | 22.6 | 28.8 | 32.5 |
| Bottom 25% | 7.9 | 15.3 | 28.5 | 48.3 |

NOTE: Includes 1,417 stations with complete data for FY 2004.

---

[6] Comparisons based on other traditional performance measures, which are not reported, are similar.

for PPM2. Indeed, nearly 20 percent (i.e., 12.7 plus 5.9 percent) of stations ranked in the highest quartile based on BOX were actually in the lowest two quartiles (i.e., the bottom half) of performance based on PPM2. In other words, a substantial proportion of highly ranked stations according to BOX were actually below-average performers according to the conceptually defensible PPM2.

A similar pattern can be seen for those stations ranked in the bottom quarter of the distribution of BOX. For example, 7.9 percent of the lowest mission-box performers are among the top quarter of stations evaluated using PPM2 and 24 percent (i.e., 7.9 plus 16.1) of those in the bottom half according to BOX are actually in the top 25 percent of stations according to PPM2. Clearly, if measures are used that do not adjust adequately for variations in recruiting difficulty, station rankings will be unreliable. Thus, we conclude that none of the traditional measures provides satisfactory approximations to our most-preferred measure, PPM2.

# CHAPTER FIVE

# Choosing Performance Windows and Organizational Units for Evaluation

In this chapter, we evaluate alternative time intervals and organizational units for performance measurement. First, we discuss and assess appropriate performance windows for which performance metrics should be calculated. Second, we turn to the organizational unit and evaluate the efficacy of measuring performance for individual recruiters relative to station-level (team) performance measurement.

## Using the Performance Window to Control for Random Outcomes

In the short run, observed production by a station's recruiters is a function of several factors, including the effort and skill of recruiters and the quality of the market in the station's territory. In addition, outcomes depend on random factors that are not captured by any of the variables used in our econometric models. Such factors may include unobserved characteristics of the potential recruits and those who influence them, such as parents, friends, and teachers. Other random events may include exposure to an effective advertisement, newly emerging personal issues, or job-related circumstances. As a result, day-to-day or even month-to-month production may be due primarily to luck, both good and bad. To the extent that random events average out over time, performance metrics computed for longer performance windows will tend to be more sensitive to cross-station variation in recruiter effort and skill and, therefore, will be more accurate for evaluating recruiter productivity.

Indeed, our previous research has demonstrated that variations in the frequency with which stations meet their regular Army missions have large components of randomness.[1] In fact, it appears that most stations that fail to make mission for several consecutive months, even up to six months, have been unsuccessful because they have either been unlucky, are located in difficult recruiting markets, or both. Only after a full year's production can one feel reasonably certain (i.e., 80 percent confident) that a recruiter's or a station's performance, even when one controls for market quality, actually reflects effort or skill differentials.

Of course, we saw in Chapter Four that mission-box accomplishment (i.e., BOX) is far from an ideal performance metric, largely because variations in mission levels do not closely reflect variations in market quality. To see whether the Dertouzos and Garber (2006) conclusions regarding the effects of lengthening performance windows hold up for measures that allow for three distinct dimensions of market quality, we assessed the extent to which enlistment outcomes are predicted more accurately for successively longer performance-evaluation periods, using the graduate-senior-other model presented in Chapter Two (equations 2.9).

In a series of regressions, we examined the degree to which the sum of monthly enlistment predictions from our model fit or explained the sum of actual enlistments over alternative periods.[2] Table 5.1 presents the results. For a single month, the model performs best for graduates, with a goodness of fit ($R^2$) of 0.318. The model for other enlistments explains less of the variance, with an $R^2$ of 0.274. The senior model has an $R^2$ of about 0.098.

---

[1]   Dertouzos and Garber (2006, pp. 99–101).

[2]   Based on our econometric estimates of the graduate, senior, and other contract models, we generated predicted levels of enlistments for each station for each month. We then regressed actual enlistments in each category on the predicted levels based on the values of exogenous variables. For multiple periods, we merely summed the actual and predicted values over the relevant number of months.

**Table 5.1**
**Percent of Across-Station Variation in Enlistments Explained by Regression Models over Successively Longer Performance Windows**

| No. of Months | Graduates | Seniors | Other |
|---|---|---|---|
| 1 | .318 | .098 | .274 |
| 2 | .415 | .153 | .416 |
| 3 | .521 | .199 | .503 |
| 6 | .652 | .311 | .653 |
| 9 | .718 | .393 | .731 |
| 12 | .758 | .462 | .776 |
| 18 | .812 | .541 | .825 |
| 24 | .836 | .590 | .849 |
| 24[a] | .963 | .915 | .970 |

[a] Includes dichotomous variable = 1 for each of over 1,500 stations.

As expected, because random outcomes tend to average out over time, the models perform increasingly well at predicting enlistment levels as longer time intervals are examined. For a performance window of six months, for example, the computed $R^2$ measures exceed 0.65 for both graduate alphas and other enlistees, while for senior alphas the model continues to perform relatively poorly, with a fit of only 0.311. The relatively poor performance of our model for seniors may reflect unobserved, market-by-market differences in the quality of high school lists, access to school populations, or variations in school-year calendars. With performance windows exceeding six months, the fits of the models continue to improve, eventually achieving $R^2$ values exceeding 0.80 for graduates and others. The fit for the senior-alpha model, even at 24 months, remains below 0.60.

The improvements in $R^2$ as longer time intervals are considered reflect the importance of randomness in determining enlistment outcomes. As the performance window lengthens, randomness becomes less important and outcomes depend increasingly on market quality and effort, as predicted by the underlying model. The remaining unexplained variation can be viewed primarily as unobserved (and unpredicted) productivity differences that could be due to varying levels of effort and skill or to local market conditions that are not captured by the variables included in the regression model. To the extent that these differences are fixed over time, adding dichotomous variables representing each of the roughly 1,500 stations in our sample—i.e., adding station-level fixed effects to the regression—can capture them. The inclusion of these "dummy" variables increases the $R^2$ statistics to 0.963, 0.915, and 0.970 for graduates, seniors, and other enlistments, respectively. Comparing the last two rows of Table 5.1 indicates that over a two-year period and for all three enlistment categories, factors specific to stations that are constant over our sample period account for between three-quarters and four-fifths of the variation left unexplained by our models.[3]

To promote both equity and productivity, performance measures should be insensitive to random events. Thus, our analysis suggests that monthly metrics do not sufficiently reflect the factors that are relevant for evaluation, namely, recruiter effort and skill. This is because much of the variation in contract production during a single month is due to factors beyond the control of recruiters. However, the importance of luck is significantly attenuated for performance windows of six months, the time interval traditionally and currently used by USAREC in aggregating production points in the awards program. Indeed, the fact that the highest award levels are feasible only through aggregate production over at least a two–year period further reduces the role of randomness. Still, the large and discontinuous number of points given to recruiters only when their stations achieve mission during a single month places too much emphasis on short-term performance. Therefore, the Army should consider lengthening the performance window.

---

[3]  For example, in the case of graduate alphas, our estimates applied to two years of data leaves 16.4 percent (i.e., 1.00 – 0.836) of the variation unexplained, and adding station fixed effects reduces this unexplained variation to 3.7 percent (i.e., 1.00 – 0.963) implying a 77.5 percent reduction in unexplained variation.

## The Use of Station Versus Individual Performance Evaluation

In the previous section, we found that randomness greatly affects station-level enlistment outcomes when performance windows are short, such as a month or a quarter. As the window lengthens, outcomes become more predictable and differences in contract production become more reliable indicators of differences in market factors and recruiter skill, effort, or both. From both equity and productivity perspectives, it makes sense, then, to focus on evaluation periods of between six months and one year. One might infer that analogous arguments apply to other performance measurement approaches that also serve to reduce the proportion of observed variation in contract production that is due to random events. For example, combining organizational units and evaluating them jointly—combining individual recruiters into stations, companies, etc.—would serve to smooth or average out the effects of random events that are beyond the control of recruiters.

However, changing the management unit of performance raises issues related to the incentives facing recruiters—in particular, how recruiter effort responds to different levels of uncertainty. In this section, we consider this issue in the context of choosing between individual and station-level performance evaluation, develop empirical implications from a theoretical model of recruiter behavior, and then test these implications. Our empirical analysis uses data from FY 1999 through FY 2001, a period during which the recruiting command switched from individual missioning and performance review to station missioning and performance evaluation. The switch was staggered, with two brigades, comprising almost 40 percent of the nation's recruiting stations, changing in FY 2000. The remaining three brigades switched to station missioning and performance evaluation the following fiscal year. This two-stage introduction of station missioning provides a unique opportunity to assess the impact of station missioning using econometric methods.

## The Impact of Station Missioning: Theory and Simulations

Team- and individual-based performance evaluation and rewards are used by many private-sector organizations.[4] But neither the academic nor the trade literature provides general guidance for choosing between team and individual goals and rewards systems. The reason for this lack of guidance may have important practical implications for the Army. In particular, whether productivity will be higher under team- or individual-based incentives depends on several, sometimes subtle, factors that are likely to vary considerably across Army recruiting stations.

In general terms, station missioning (the Army version of team based selling) has a major potential advantage, namely encouraging teamwork in a task for which cooperation has considerable scope for improving productivity. For example, one recruiter can pitch in when another is overwhelmed or sick; a corporal may concentrate on selling recruits, with an older recruiter helping out when it is time to sell the parents; and recruiters may take advantage of specialization of labor by having recruiters spend more time doing what they do best (e.g., face-to-face versus telephone selling). But, depending on the personalities, histories, and interpersonal dynamics among recruiters in a station, using station missioning can undermine productivity. For example, some team members may tend to shirk ("free ride" in the economic jargon), in which case other recruiters might be resentful, and morale and effort could suffer across the board.

On the surface, evidence presented earlier suggests that team performance evaluation may be more advantageous in the Army context. To begin with, we have seen that much of the month-to-month variation in contract production by recruiters is due to events that are beyond the recruiters' control. Although monthly enlistment flows at the station level remain highly uncertain, it is clear that the uncertainty is lower at the station level for stations with more than one recruiter. This may be desirable, especially from a fairness perspective.

In order to evaluate the productivity or efficiency implications of the risk-reducing aspect of teaming, we developed an analytic model

---

[4]   We know of no quantitative information about how often organizations use team versus individual approaches.

based on labor economics and the theory of time allocation.[5] In this model, recruiters allocate their time between leisure and effort on the job, and more effort increases the expected number of enlistments. The recruiter values both leisure and enlistments. Expected enlistments are valued because they increase the probability of achieving the mission, for which there are valued rewards such as bonus points toward command-level awards (stars, badges, rings), better performance reviews and promotion prospects, and lower likelihoods of negative outcomes such as closer supervision and pressure from teammates. The probability of making mission depends on the effort expended and the level of the mission, as well as exogenous supply factors, such as propensity to enlist, the state of the local economy, and levels of recruiting resources, such as advertising or enlistment bonuses.

The model permits a simulation analysis of policy options concerning the allocation and level of missions as well as policies that alter the degree of randomness of outcomes for a given level of effort.[6] For example, team missions can be viewed as a policy that decreases the variance of monthly recruiting outcomes because random variations across recruiters tend to average out when summed to the station level. We also examine the length of the performance window. Should the mission be a monthly, quarterly, or annual target? Finally, the mission-box categories could be broadened or narrowed. For example, the cate-

---

[5]    We provide details in Appendix B.

[6]    The implications of the model offered in Appendix B are analyzed using numerical simulations because that model is too complicated to yield analytic solutions. A major factor contributing to the complexity of the model is the discontinuity in recruiter rewards associated with making mission. For the purposes of the current section—i.e., exploring the implications of different degrees of uncertainty facing recruiters, depending on whether they are evaluated individually or as stations (or teams)—as emphasized in our discussion of intuition, we believe that discontinuities associated with making mission are important. In contrast, the model of recruiter choice offered and analyzed in Appendix A is simpler in that it involves no discontinuity. On the other hand, it is more complicated in that there are three different enlistment categories. However, it does not require numerical simulation to explore implications and allows mathematical derivation of explicit solutions for the effort-per-recruiter functions. This derivation is required to yield explicit expressions for the intercepts of the effort equations that are needed to resolve the ambiguity of the units in which various unobservable constructs are measured.

gories formerly included distinctions between men and women, graduates and seniors, and prospects with and without prior service, further distinguished by three AFQT categories. Thus, there were over a dozen categories in the mission box. The randomness of making mission is much smaller today, because the mission box is defined over only three relatively broad categories (i.e., I–IIIA graduates, I–IIIA seniors, and all others).

The results from one set of our simulations are summarized in Figure 5.1. The model was calibrated so that the initial outcomes appeared to be realistic. The height of the bars indicates production per recruiter. The horizontal axis indicates the mission per recruiter. Simulations are performed for situations in which missions and rewards are based on individual performance and, alternatively, for teams of three recruiters.

The simulations suggest that there are gains to increasing mission at low levels of difficulty (which here correspond to low missions per

**Figure 5.1**
**Impact of Mission and Team Size on Productivity per Recruiter**

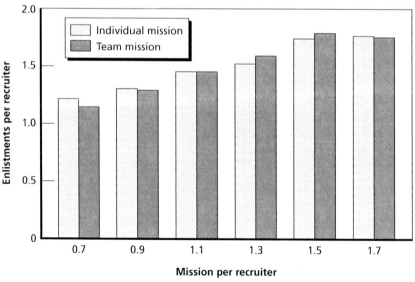

recruiter), but that these gains diminish with increasing difficulty. In fact, beyond a certain point (an average of 1.8 for the team and 1.7 for individual missions), the returns to increasing missions are eliminated and even negative. The advantage of team relative to individual incentives—i.e., the relative heights of the two columns corresponding to the mission per recruiter—varies with the level of mission difficulty. For relatively easy missions (i.e., missions of 0.7 per recruiter or lower), the individual approach dominates. However, for mission levels of between 1.3 and 1.7 per recruiter, the team approach (i.e., station missioning) leads to more enlistment per recruiter. As the mission per recruiter increases beyond 1.7, the individual performance approach again produces more enlistments.

The relevance of these theoretical patterns requires empirical validation and, if valid, their magnitudes require estimation. But some important policy implications of the theory make good intuitive sense. First, higher missions do stimulate effort, but only up to a point, which is why we specify quadratic functional forms that allow for effort to fall with difficulty in our empirical contract-production equations such as (2.9). Second, the optimal mission depends on the organizational approach and the preferred organizational approach depends on the level of the mission. For example, when missions are especially easy or difficult, individual missions are predicted to be better. Intuitively, this is because when missions are easy but outcomes are highly variable, extra effort on the part of an individual can be viewed as an insurance policy against not making mission. The extra variance associated with individual missions thus gives the recruiter incentives to work harder to reduce the risk of failure. In contrast, when missions are very high relative to what the market can be expected to allow, additional effort can be viewed as a lottery ticket. On average, recruiters will not make mission, and the likelihood that a team will achieve its enlistment goal is low. In this situation, effort is reduced by station missioning because the higher variance associated with individual missions creates a better chance that a single recruiter will get lucky, and this chance can stimulate effort. Finally, in the middle ground when missions are challenging but attainable, larger teams increase the expected payoff to extra effort.

As the team approaches the goal, collective effort is likely to result in making mission and gaining the reward for doing so.

Clearly, this model abstracts from other issues relevant to the choice of individual and team missions. For example, there may be gains to specialization or other economies of teamwork that tend to favor a team approach. Also, incentives may differ when there are teams. With large teams, some recruiters may be tempted to shirk (free ride on the efforts of their teammates). Moreover, members of smaller teams might find it easier to monitor each other's effort. In addition, team members may exert productivity-enhancing pressure, particularly to the extent that effort by teammates is observable. Thus, there may be gains or losses to teaming due to factors not modeled.

Ultimately, the choice of the best policy depends on the magnitude of tradeoffs that cannot be determined theoretically. This remains an empirical question to which we now turn.

## Empirical Evidence on the Efficacy of Station Missioning

Fortunately, the sequential adoption of station missioning in fiscal years 2000 and 2001 provides a natural experiment for evaluating the impact on production of high-quality enlistments. To do so, we gathered information for the 1999–2001 fiscal years and estimated two generalized versions of the high-quality contract production model of Dertouzos and Garber (2006), which is described and discussed in Chapter Two.

For this analysis, we generalized the high quality contract model of Dertouzos and Garber (2006, Chapter Four) by replacing (2.4′) in Chapter Two with the following more general expression for effort per recruiter in a particular month:

$$e_{ist} = 1 + (\beta + \beta_1 R_{st})d_{ist} + (\gamma + \gamma_1 R_{st})(d_{ist})^2 + (\alpha + \alpha_1 d_{ist}$$
$$+ \alpha_2 (d_{ist})^2)S_{st} , \tag{5.1}$$

where the subscripts $s$ and $t$ index stations and months, respectively and

$S_{st}$ is a dichotomous variable equaling one if station $s$ was under station missioning during month $t$ and zero otherwise,

$R_{st}$ is the recent performance ratio for month $t$, which was used in previous models (i.e., contracts divided by mission for the previous year, lagged one quarter), and

$d_{ist}$ is high-quality mission difficulty in month $t$ as defined and used previously (i.e., high-quality mission per OPRA recruiter in month $t$ divided by the marginal productivity of effort).

The two generalizations of the Dertouzos and Garber (2006) model used to examine the efficacy of station missioning differ according to the qualitative nature of effects of station missioning on (high-quality) contract production. Specifically, Model 1 allows station missioning to either increase or decrease effort per recruiter by only a constant amount to be estimated. In contrast, Model 2 is more general than Model 1, with the former allowing—as in our model whose implications are summarized in Figure 5.1—the effects of station missioning on effort to depend on the difficulty of making mission. More formally, in Model 1 we assume that both $\alpha_1$ and $\alpha_2$ in (5.1) equal zero; and we impose these values in estimation; thus, the effect of station missioning on effort-per-recruiter simply equals $\alpha$. In Model 2, we constrain none of the parameters of (5.1) in estimation, thus allowing station missioning (the variable $S_{st}$) to shift contract production by a constant amount $\alpha$ (as in Model 1) and also through changing the sensitivity of effort to mission difficulty ($d_{ist}$) and its square (i.e., allowing both $\alpha_1$ and $\alpha_2$ to differ from zero. The estimates of the parameters of (5.1) for Models 1 and 2 are presented in Table 5.2.

For Model 1, the estimated parameter representing the average impact of station missioning ($\alpha$) was positive, at 0.2005, and highly statistically significant. All things being equal, this estimated effect of station missioning on effort per recruiter implies about an 8 percent increase in high-quality enlistments for an average station that adopted station missioning during this period.[7]

---

[7]   We also estimated this model using data expressed in terms of year-to-year differences as well as with station-level fixed effects. These alternative specifications would tend to reduce

**Table 5.2**
**Effects of Station Missioning:**
**Estimated Parameters of Effort-**
**per-Recruiter Function**
**(standard errors in parentheses)**

| Coefficient | Model 1 | Model 2 |
|---|---|---|
| $\beta$ | 0.1488 (0.0131) | 0.0703 (0.0167) |
| $\beta_1$ | 0.1761 (0.0150) | 0.2120 (0.0169) |
| $\gamma$ | 0.0038 (0.0021) | 0.0116 (0.0030) |
| $\gamma_1$ | −.0088 (0.0029) | −0.0182 (0.0040) |
| $\alpha$ | 0.2005 (0.0178) | −0.2157 (0.0362) |
| $\alpha_1$ | $0^a$ | 0.1802 (0.0210) |
| $\alpha_2$ | $0^a$ | −0.0157 (0.0030) |

[a] Value imposed in estimation.

The results for Model 2—for which the effect of station missioning is also allowed to depend on the level of mission difficulty—are difficult to appreciate from parameter estimates because they involve three parameters, namely, $\alpha$, $\alpha_1$, and $\alpha_2$, the estimates of which (Table 5.2), are all highly statistically significant. The implications of these estimates are most easily understood by examining the simulation results shown in Table 5.3, which reports for various levels of mission difficulty the proportionate change in high-quality contracts.

As can be seen in Table 5.3, when mission difficulty is at the lowest (easiest) level listed in the table (difficulty = 0.5), production

---

any biases due to correlation between unobserved station-level characteristics or seasonal events that affected which stations first moved to station missioning and when these changes occurred. The results concerning the average impact of station missioning on high-quality contract production were virtually identical for these versions of the model to those reported in Table 5.2.

**Table 5.3**
**Mission Difficulty and the Impact of Station Missioning on High-Quality Contract Production**

| Mission Difficulty | Proportionate Change in High-Quality Contracts Due to Station Missioning |
|---|---|
| 0.5 | −0.1295 |
| 1.0 | −0.0512 |
| 1.5 | 0.0193 |
| 2.0 | 0.0819 |
| 2.5 | 0.1367 |
| 3.0 | 0.1836 |
| 5.0 | 0.2928 |
| 7.0 | 0.2764 |
| 9.0 | 0.1344 |
| 11.0 | −0.1332 |

NOTE: Mission difficulty is the high-quality mission per OPRA recruiter divided by the marginal productivity of effort ($c_S^{\cdot}$). A majority of stations confronted difficulty levels between 2 and 3.

is estimated to be almost 13 percent lower under station missioning than under individual missioning. When mission difficulty is in the middle range, however, production is substantially higher when station missions are used. For example, for difficulty = 2.5 (near where many of our observations fall) and difficulty = 5 (which is far from typical in our sample), station missioning is estimated to increase high-quality contracts by almost 14 and more than 29 percent, respectively. At the highest level of difficulty in the table (difficulty = 11), making mission is almost impossible, and adopting station missioning reduces effort and contract production. During this period, about 10 percent of the stations were in the tails of the difficulty distribution—i.e., values for which individual missioning increases production—suggesting that, while introducing station missioning increased productivity overall or on average, it actually reduced recruiting productivity for a small subset of stations. In sum, the estimates reported in Table 5.3 are consistent with the hypothesis or prediction discussed above—namely, "when missions are especially easy or especially difficult, individual missions are predicted to be better"—that was generated by simulating the model of recruiter choice detailed in Appendix B.

We emphasize that, although station missioning appears to have increased high-quality contract production during FYs 1999–2001, if recruiting were to become either much easier or much more difficult (given mission levels) than during that period, this basic conclusion could change. For example, with significant changes in the mission level, enlistment propensity, or economic conditions, it could become effort-enhancing to return to individual missioning. It could make the most sense to adopt a flexible performance measure that requires individual achievement when the unit makes mission and also allows for individual recruiters to succeed, even when the station fails.[8]

---

[8]   The Army does require some individual production by a recruiter to earn bonus points for his or her station (or a higher management unit such as a company or battalion) making mission. In addition, production points have, in the past, generally been awarded for individual production even when the station fails. However, our estimates suggest that further modification of Army policies might be desirable so that individuals continue to be motivated under extreme circumstances.

# CHAPTER SIX

# Conclusions

## Summary of Results

Performance metrics are the benchmarks by which individuals and management units of an organization are evaluated. If designed effectively, such measures can serve to motivate personnel and their managers and help ensure that individual incentives are well aligned with those of the organization. For Army recruiting, an ideal performance metric would isolate true productivity—a combination of effort exerted and skill applied—by making adjustments based on several factors. In particular, an effective performance metric for recruiters, stations, and other management units, should do the following:

- Adjust for exogenous factors, such as the quality of local markets or regions based on economic conditions or demographics.
- Account for differences in enlistment propensity over time or among local markets or regions.
- Consider differences in the relative difficulty of recruiting population subgroups, such as graduates and seniors with high test scores.
- Reflect the value the organization places on separate enlistment categories.
- Account for the random nature of observed outcomes in choosing units for evaluation (organizational units or performance) periods.

As we have seen, five common ("traditional") measures of recruiting performance all fail in one or more of these crucial dimensions. For example, the write rates and numbers of high-quality contracts per recruiter both fail to consider market-to-market variations in the difficulty of recruiting. In addition, these measures implicitly make extreme assumptions about the relative difficulty—as well as the relative value to the Army—of producing contracts of various types. For example, the write rate gives equal weight to all categories by simply adding up counts of all contract types. In contrast, measures that consider only high-quality enlistments give no weight at all to lower-quality ones.

In our econometric analysis of high-quality seniors, high-quality graduates, and other enlistees, we found significant differences across stations' market areas in the relative difficulty of enlisting youths in these subpopulations. Performance measures should reflect these differences. We also found significant differences in the relative difficulty of enlisting males and females. To the extent that the Army differentially values these and other relevant population segments (for example, by finer AFQT, education, or MOS distinctions), an ideal PPM should reflect these differences.

Since missions are allocated to battalions and stations based to some extent on local characteristics that contribute to market quality,[1] achieving mission box—another traditional performance measure—adjusts somewhat for the difficulty of making mission, at least for the broad missioned categories. However, these adjustments are far from perfect because they do not adequately consider relevant demographic and economic characteristics, nor do they account for market-by-market differences in the ease of recruiting distinct population segments. Perhaps most importantly, making monthly mission box at the station level is subject to a great deal of randomness. That fact, along with the discontinuous and significant reward (award points) given to recruiters in those stations that make mission, means that rankings

---

[1]  Recently, the missions have been based on the "50/50 model" which apportions mission based on past (previous 36 months) Army and other service enlistments. This method implicitly accounts for market difficulty but falls short of being ideal because it does not adequately adjust for differences in effort levels and recent economic or demographic changes that could alter the difficulty of making mission.

based on the frequency of making mission box (even over a two-year period) are not fully reliable indicators of performance.

In contrast to the traditional measures of performance, we derived a preferred performance metric that has all the desirable characteristics listed at the beginning of this chapter. We used a model of enlistment supply that considers the joint roles of market characteristics and recruiter effort in the face of allocated missions. Using monthly, station-level data from FYs 2001–2004, we estimated the parameters of this model. Using these estimates, we then computed the PPM, compiled station rankings based on performance over the last year of the sample, and compared these rankings with those based on traditional measures.

We found limited correlations between traditional measures and the PPM, suggesting that rankings based on common metrics are likely to be misleading about recruiter productivity. To illustrate this point, we compared station rankings, by quartile, for the frequency of making mission box (number of months during FY 2004) with rankings based on the PPM. We found that the frequency of making mission box is a poor indicator of actual productivity, as measured by the PPM. Indeed, of those performing in the top 25 percent as measured by making mission box, only 46 percent were in the highest rank (top 25 percent) based on the PPM. At the other extreme, many stations that consistently failed to make mission were ranked as top-quarter performers according to the PPM.

We also examined the appropriate time unit for analyzing performance. For all measures, randomness is a dominant factor in determining *monthly* outcomes for small managerial units, such as individual recruiters or stations. Only after a period of time ranging from three to more than twelve months, depending on the contract type, do enlistment outcomes appear to reflect primarily systematic differences in recruiter productivity rather than good or bad fortune. This randomness is a larger issue in the current way the Army implements the recruiter reward system in which a discontinuous and relatively large number of bonus points are awarded for making monthly mission. Because of the small monthly target numbers, randomness can dominate the frequency of making mission over long periods of time, resulting in large discrepancies between productivity and success as measured by accumulated award points.

However, when missions are either very easy or very difficult, valid reasons for maintaining higher variance in performance measures remain. This is because when missions are unusually easy, recruiters will attempt to insure themselves against a low probability of failure by working harder. In contrast, recruiters are willing to work harder when missions are very difficult because a high variance gives them some chance of success (like buying a lottery ticket). Econometric estimates indicate that under most circumstances prevailing during FYs 2000–2001, the effect of team or station missioning (as opposed to individual missioning) was positive, increasing high-quality contracts by an average of 8 percent. It is noteworthy that stations where the effect was negative—those at either extreme of the "mission difficulty" continuum—were given missions that were exceptionally easy or exceptionally hard to achieve in light of local market conditions. In sum, a better method of allocating missions might have eliminated the positive influence of greater uncertainty in extreme cases.

## Implications for Policy

Based on our findings, we believe that the Army should adopt modest and perhaps gradual reforms in the way recruiters are currently evaluated. Although common performance metrics are clearly flawed, four major caveats make us reluctant to recommend an immediate and wholesale adoption of our preferred performance metric.

1.  The current missioning process and associated awards (such as badges, stars, and rings) are a deeply ingrained part of the Army recruiting culture. Despite its flaws, the system has, for the most part, succeeded in attracting high-quality youth to enlist. Sudden dramatic changes are likely to meet resistance that could undermine recruiter morale and productivity at a time when declines in enlistments would be extremely problematic.

2. The PPM, in comparison to common measures, is conceptually more complex and requires fairly sophisticated econometric analysis for implementation. Although the PPM could be implemented by USAREC, doing so would place an additional burden on USAREC. Perhaps more important, the new metric would not be transparent and intuitive to personnel in the field. Although we believe that a new performance metric would be more equitable, perceptions are sometimes more important than reality.

3. The details of the PPM depend on assumptions about enlistment supply relationships, the nature of recruiter behavior in response to the current mission system, and the relative value of enlistments in different segments of the target population. Additional analyses should be conducted before settling on a particular version of the PPM.

4. Our research focused primarily on three markets: seniors in AFQT categories I–IIIA, graduates in I–IIIA, and all other contracts. Our exploratory work distinguishing males and females indicates that substantial distinctions exist between these segments—distinctions that should also be considered in the design of a performance metric. Indeed, there are likely to be other segments—based on education or MOS preferences, for example—that are worth considering.

Despite these caveats, we recommend that USAREC consider making some short-term adjustments to the current approach to evaluating recruiters. In particular, USAREC should do the following:

• Improve mission allocation algorithms to reflect variations in market quality and differences in market segments.

Current mission allocations do not do a very good job of adjusting for station-level variation in crucial economic and demographic factors that affect the productivity of recruiter effort and skill. Even measures that evaluate performance relative to mission are inadequate, primarily because the methods used to determine mission levels do not accurately reflect market quality.

- Lengthen the performance window to at least six months or smooth monthly rewards by decreasing the emphasis on monthly mission accomplishment.

We have seen that until at least six months of production are observed, recruiting outcomes at the station level can be dominated by randomness in the enlistment process. The consequences of this randomness are exacerbated by a system of evaluation and awards that provides a discontinuous and significant reward for making mission in a single month. A good portion of recruiters who have consistently achieved mission are, in reality, not much more productive than their less-successful counterparts who have suffered from bad timing (contracts not written when their monthly mission demands them), bad luck, or both.

- Consider a more refined system of rewards for additional enlistment categories such as males, higher education or AFQT levels, critical MOSs, or those willing to accept longer terms of service.

Our exploratory research distinguishing males and females strongly suggests that additional market segments may differ significantly in terms of their recruiting difficulty. These differences probably vary systematically on a market-by-market basis. It would not be advantageous to allocate missions for detailed subcategories—especially if mission accomplishment in a single month or even a few months is associated with a large bonus award—because this would further increase the importance of randomness in determining mission success. A supplemental system of allocated points based on the distribution of enlistments among categories of special interest to USAREC would better reflect true productivity as well as providing additional incentives for recruiters to meet overall Army objectives.

- To minimize resistance, include education and outreach when implementing reforms.

Organizational change is always difficult, especially when there are perceived winners and losers. Modest efforts to explain and justify the changes, including the technical nature of the methods that are at their foundation, could increase acceptance. If the performance measures are perceived as fair, with every station given a reasonable chance of success if the recruiters work hard, resistance will be reduced.

Although it is impossible to predict the efficiency gains that could emerge from such reforms, they are likely to dwarf the costs of implementing them. As demonstrated in previous research (Dertouzos and Garber, 2006, Chapter Six), better mission allocations during the 2001–2003 period could have improved average productivity by nearly 3 percent. Much of these gains were due to an increased willingness on the part of stations that had a previous history of success (by conventional measures) to do more for the Army. There is good reason to believe that a revised performance metric that reflects Army values and accurately assesses recruiters' hard work and talent in meeting organizational objectives would also have significant benefits.

# Allocation of Recruiter Effort: Implications of a Microeconomic Model

In this appendix, we use microeconomic theory to derive effort-per-recruiter equations for the three contract types that were missioned during our analysis period, which are given by equation (2.8). These expressions allow us to analyze two issues of fundamental importance for our empirical analysis and interpretation of the results. First, the analysis enables us to resolve the ambiguities associated with the scales (or, equivalently, the units of measurement) of unobservable variables such as recruiter effort and skill and their marginal productivity. As described in Chapter Four, resolving this ambiguity is necessary to derive performance measures focused on effort and skill. Second, we use the effort-per-recruiter equations derived in this appendix to show that—while recruiters with better skills will tend to expend less effort than less-skilled recruiters, other things being equal—recruiters with better skills will nonetheless be more productive. As we prove in this appendix, this is because, as skill levels increase, effort levels fall more slowly than skill levels rise. Thus, effort plus skill increases with skill levels.

## A Model of Recruiter Choice of Effort

Assume that the utility function of the representative recruiter in station $s$ is given by:

$$U_{is} = \sum_j \{\pi_j + \alpha_{1j} \frac{g_{sj}}{N_s c_{sj}^*}\} \ln(Ec_{sj} / N_s) - W\left(\sum_j e_{sj}\right) \quad (A.1)$$

where:

$U_{is}$ is the utility of the representative on-production recruiter in station $s$,

the constants $\pi_j$ ($j = G, S, O$) are known and most naturally interpreted as the point values in the Army recruiter award program for producing contracts of type $j$,

$\sum_j e_{sj} = e_s$ is the total effort expended by recruiters in station $s$ to recruit youth of all types, and

the function $W$ is the disutility of (or disincentive for) expending effort and has positive first and second derivatives.[1]

We assume that a representative recruiter chooses effort levels to allocate to producing contracts of each of the three types, namely, $\{e_{sj}\} = \{e_{sG}, e_{sS}, e_{sO}\}$, to maximize his or her utility.

Our reasons for expressing recruiter utility as in (A.1) are as follows. First, (A.1) expresses recruiter utility as an increasing function of the expected value of each contract type produced and a decreasing function of total effort expended across contract types. Second, the term in {} in (A.1) expresses the idea that recruiters value contracts for two reasons: (1) the points awarded for each contract (the $\pi_j$ for $j = G, S, O$), and (2) progress toward meeting the recruiting goal (i.e., $g_{sj}$). Regarding the latter, as expressed in (A.1), we assume that contracts of a particular type are more highly valued by recruiters—other things being equal—the more difficult the station's goal is for that type (i.e., $g_{sj} / N_s c_{sj}^*$). Third, we choose the particular form in which expected contract levels enter the utility function (i.e., the logarithmic form) for analytic tractability. In particular, using this functional form enables us to derive explicit solutions for the effort equations and

---

[1]  Thus, we are assuming that the disutility of expending effort increases with effort at an increasing rate.

thereby to implicitly express the effort and skill levels pertinent to the three contract types in the same (or "common") units.

## The Effort-per-Recruiter Equations

Next we derive equations determining the per-recruiter levels of effort allocated to trying to enlist youth of the three types. The relative intercepts (i.e., ratios of all three pairs of these intercepts) of the equations we derive are *known* constants, and, as discussed in Chapter Two, this property of the effort-per-recruiter equations is central to developing and implementing conceptually grounded performance metrics that combine counts of produced contracts of various types. More specifically, knowing the relative intercepts of the effort-per-recruiter equations—and imposing these relative values in estimating contract-production equations (2.9)—enables us to implicitly pin down the scales of the unobservable marginal productivity of effort $\{c^*_{sj}\}$, effort levels, and skill levels and (crucially) to *express them in common units* across the three contract types. With these unobservable variables expressed in common units, it is then meaningful to sum over contract types the contracts actually produced, weighted (or multiplied) by estimated station-specific levels of difficulty of recruiting youths of those types— as we do in constructing our preferred performance metrics presented in equation (2.10). Our PPMs are those sums for a particular station during a particular performance-evaluation time interval (or performance window) and can be interpreted as estimates of the effort plus skill applied by each recruiter in the station to produce the contracts actually produced.

The necessary (or first-order) conditions for an interior solution to the optimization problem (i.e., a solution with positive values for each of the three effort levels), which we denote by $\{e^*_{sG}, e^*_{sS}, e^*_{sO}\}$ are

$$\frac{\partial U}{\partial e_{sj}} = \{\pi_j + \alpha_{1j} \frac{g_{sj}}{N_s c^*_{sj}}\} \frac{\partial \ln(Ec_{sj})}{\partial Ec_{sj}} \frac{\partial Ec_{sj}}{\partial e_{sj}} - W'(e^*_s) \frac{\partial e_s}{\partial e_{sj}} = 0 ,$$

which implies that

$$\{\pi_j + \alpha_{1j} \frac{g_{sj}}{Nc_{sj}^*}\} \frac{c_{sj}^*}{Ec_{sj}} = W'(e_s^*).$$ (A.2)

Rearranging the contract production equations (2.7)

$$Ec_{sj} = c_{sj}^*(e_{sj} + v_{sj}) \text{ for } j = G, S, O$$

yields

$$(e_{sj} + v_{sj}) = \frac{Ec_{sj}}{c_{sj}^*},$$ (A.3)

and combining this expression with (A.2) and rearranging yields the effort equations

$$e_{sj}^* = \{\frac{\pi_j}{W'(e_s^*)} + \alpha_{1j} \frac{g_{sj}}{N_s c_{sj}^* W'(e_s^*)}\} - v_{sj} \text{ for } j = G, S, O.$$ (A.4)

Equations (A.4) indicate that (1) recruiters in stations with higher average skill levels will exert less effort, other things being equal; (2) assuming that skill levels are uncorrelated with goals and the marginal productivity of effort,[2] we can treat skill as part of the disturbance terms in the contract-production equations; and (3) most important, the intercepts of the effort equations are $\pi_j / W'(e_s^*)$ for $j = G, S, O$. Thus, the microeconomic model implies that the intercepts of the effort equations are proportional to award points. In estimating our contract-production equations, we set the value of the intercept of the effort equation for each contract type equal to the point value of that con-

---

[2]  This assumption would be violated if recruiter ability could be predicted in advance, and recruiters could be given missions reflecting ability differences and/or assigned to markets with systematically different quality levels. We have seen no evidence that this is the case.

tract, which implicitly expresses all effort (and skill) levels in common units that can be summed across contract types.

In fact, the effort-per-recruiter equations that we specify and estimate are more general than those given by (A.4). In particular, we estimate effort-per-recruiter equations of the form

$$e_{isj} = \pi_j + (\beta_j + \beta_{jj}R_{sj})d_{sj} + (\gamma_j + \gamma_{jj}R_{sj})d_{sj}^2 \text{ for } j = G, S, O, \quad \text{(A.5)}$$

where $d_{sj} = g_{sj} / N_s c_{sj}^*$ is the difficulty of making the mission for contract type $j$, and $R_{sj}$ is a measure of the station's recent past success in recruiting youths of type $j$. Thus, (2.8) generalizes (A.4) by adding terms involving difficulty squared and recent past success. Note, however, that the expression in {} in (A.4) equals $1 / W'(e_s^*)$ times the expression in {} in the recruiter's utility function given by (A.1). The reason that this occurs is that none of the terms in {} in the utility function is endogenous—which implies that all the terms in {} in (A.1) are constant with respect to the chosen effort levels. Thus, maintaining a utility function of the form (A.1), if we were to replace the expression in {} in (A.1) with *any* expression that is invariant to the choice of effort levels, we would obtain an expression for effort-per-recruiter in the form of (A.4) with the term in {} in (A.4) merely being replaced by $1 / W'(e_s^*)$ times the revised expression in {} in the recruiter's utility function. Thus, using (2.8) as our empirical specification for effort-per-recruiter is fully consistent with the theoretical apparatus analyzed in this appendix.[3]

---

[3]   More formally, applying the steps we use to derive (A.4) to a generalization of the utility function (A.1) given by

$$U_i = \sum_j \{\pi_j + (\alpha_{1j} + \alpha_{2j}R_j)\frac{g_{sj}}{N_s c_{sj}^*} + (\gamma_{1j} + \gamma_{2j}R_j)(\frac{g_{sj}}{N_s c_{sj}^*})^2\}\ln(Ec_{sj} / N_s) - W\left(\sum_j e_{sj}\right)$$

leads to an effort-per-recruiter equation of the form (2.8), which is the form used in our empirical analyses.

## Derivation of the Perferred Performance Metric

Regarding performance measurement, equations (A.4) also indicate that it is not possible, given our models and available data, to estimate effort per recruiter at the station level. This is because, as discussed in Chapter Two, we have no means for empirically distinguishing between the effects of effort and skill on contract production.

Our PPM is derived by summing equations (A.3) over contract types:

$$\sum_j (e_{sj} + v_{sj}) = \sum_j \frac{Ec_{sj}}{c_{sj}^*}. \tag{A.6}$$

Since expected contracts are not observable, to compute performance measures we replace expected contracts with the actual numbers of contracts signed to produce an estimate of the total effort plus skill—across recruiters and contract types—required to produce the observed contracts (i.e., the $c_{sj}$) and express this on a per-recruiter basis to produce a station-level performance measure that is comparable across stations with different numbers of recruiters:

$$PPM = \frac{1}{N_s} \sum_j \frac{c_{sj}}{c_{sj}^*} = \frac{1}{N_s} [ \frac{c_{sG}}{c_{sG}^*} + \frac{c_{sS}}{c_{sS}^*} + \frac{c_{sO}}{c_{sO}^*} ], \tag{A.7}$$

which is the form presented in (2.10) in Chapter Two.

## How Are Effort and Skill Levels Related?

In closing, we show that the effort-per-recruiter equations (A.4) imply that (1) recruiters with better skills will, other things being equal, expend less effort than less-skilled recruiters (i.e., $\partial e_{sj}^* / \partial v_{sj} < 0$), but (2) effort falls more slowly than skill increases (i.e., $\partial e_{sj}^* / \partial v_{sj} > -1$). Thus, as skill increases, *effort plus skill* increases (because effort falls more slowly than skill rises), and, in light of our contract production equations,

$$Ec_{sj}^* = c_{sj}^*(e_{sj}^* + v_{sj}) \text{ for } j = G, S, O, \tag{A.8}$$

more highly skilled recruiters will be more productive (i.e., they have higher levels of expected contract production).

To show that (A.4) implies $-1 < \partial e_{sj}^* / \partial v_{sj} < 0$, we rewrite (A.4) as

$$e_{sj}^* = \frac{1}{W'(e_s^*)}[\pi_j + \alpha_{1j}\frac{g_{sj}}{N_s c_{sj}^*}] - v_{sj} \tag{A.4'}$$

and note that

$$[\pi_j + \alpha_{1j}\frac{g_{sj}}{N_s c_{sj}^*}]$$

is exogenous—in particular, this expression does not depend on skill or effort. Then, differentiating (A.4') with respect to skill using the chain rule yields

$$\frac{\partial e_{sj}^*}{\partial v_{sj}} = [\pi_j + \alpha_{1j}\frac{g_{sj}}{N_s c_{sj}^*}]\frac{d\left(\frac{1}{W'(e_s^*)}\right)}{dW'(e_s^*)}\frac{dW'(e_s^*)}{de_s^*}\frac{\partial e_s^*}{\partial e_{sj}^*}\frac{\partial e_{sj}^*}{\partial v_{sj}} - 1. \tag{A.9}$$

Rewriting (A.9) by substituting

$$(1) \quad \frac{d\left(\frac{1}{W'(e_s^*)}\right)}{dW'(e_s^*)} = \frac{-1}{[W'(e_s^*)]^2},$$

$$(2) \quad \frac{dW'(e_s^*)}{de_s^*} = W''(e_s^*), \text{ and}$$

(3) $\dfrac{\partial e_s^*}{\partial e_{sj}^*} = 1$,

and collecting terms yields

$$\left( 1 + \frac{[\pi_j + \alpha_{1j}\dfrac{g_{sj}}{N_s c_{sj}^*}]W''(e_s^*)}{[W'(e_s^*)]^2} \right) \frac{\partial e_{sj}^*}{\partial v_{sj}} = -1, \qquad \text{(A.10)}$$

and, since $\dfrac{[\pi_j + \alpha_{1j}\dfrac{g_{sj}}{N_s c_{sj}^*}]W''(e_s^*)}{[W'(e_s^*)]^2} > 0$,

it follows from inspection of (A.10) that

$$-1 < \partial e_{sj}^* / \partial v_{sj} < 0. \qquad \text{(A.11)}$$

# Recruiter Behavior in the Face of Risk

In this appendix,[1] we present a model of recruiter behavior that explores the relationship among recruiter effort, mission difficulty, and the level of uncertainty associated with attaining an enlistment target. The model parameters used in this appendix have not been calibrated to reflect real-world outcomes. Instead, the model is designed to illustrate the direction of plausible impacts that could occur as the Army chooses policies that either change the level of the mission, alter the uncertainty of making that mission, or both.

## A Model of Recruiter Behavior

We assume that the key decision facing a recruiter is to determine the amount of effort to devote to recruiting. The recruiter expends effort based on the incentives and rewards for effort and the disutility or costs of effort. A recruiter receives a fixed reward for production at or above a particular threshold (the mission). The recruiter's decision is complicated by the fact that the recruiter faces an uncertain supply of potential recruits. Therefore, achieving mission depends both on the level of effort and the uncertain supply of recruits (randomness). Assuming that the recruiter is risk neutral, the recruiter's problem can be mathematically represented as follows:

$$\underset{e}{Max} \quad E[I(e)] - C(e), \tag{B.1}$$

---

[1] The analysis reported in this appendix was conducted by Neeraj Sood.

subject to the constraints

$$y = f(e, \varepsilon) \text{ and} \tag{B.2}$$

$$I_t = \begin{matrix} I_0 & \text{if } y < \overline{y} \\ I_0 + M & \text{if } y \geq \overline{y} \end{matrix} . \tag{B.3}$$

Equation (B.1) states that the recruiter chooses his or her effort level to maximize expected income less the disutility or cost of exerting effort. Equation (B.2) is the production function, where $y$ is the number of recruits signed, $e$ is the effort and $\varepsilon$ is a random component reflecting the uncertain supply of potential recruits. Equation (B.3) is the incentives structure, where $\overline{y}$ is the mission, $M$ is the reward for making mission, and $I_0$ is the recruiter's basic compensation.

The problem is intractable in this general form, and we need to impose some simplifying assumptions to develop insights about recruiter behavior. Thus, we assume a very simple form for the production function. In particular, the number of recruits signed is the sum of effort exerted and a random component reflecting uncertain supply, with the random component assumed to be distributed normally with a mean of zero and standard deviation treated as an exogenous parameter. Formally, the production function is

$$y = e + \varepsilon, \quad \varepsilon \sim N(0, \sigma), \tag{B.4}$$

where $\sigma$ is the standard deviation of the random term, with higher values of $\sigma$ meaning that recruiters face more uncertainty about the supply of youths willing to enlist during the current period.

The first order condition for maximizing utility can now be expressed as

$$M \phi \frac{\overline{y} - e}{\sigma} \frac{1}{\sigma} = C'(e), \tag{B.5}$$

where $\phi(.)$ is the density function of a standard normal distribution. The left-hand side of equation (B.5) is the marginal increase in expected compensation for an additional unit of effort, and the right hand side is the marginal cost of exerting an extra unit of effort. This means that the recruiter will increase effort up to the point where the marginal benefit from increasing effort equals the marginal cost. If the recruiter exerts effort beyond this point, his or her expected utility will decline.

## Policy Options

We analyze in general terms the effects of two options the Army could consider for increasing recruiter effort. The first option is to increase missions. The second option is to reduce the level of uncertainty. Under most circumstances, moving from individual to station-level missions or using performance measures that average performance over more months will reduce the level of uncertainty.

Given equation (B.5), we use standard comparative-static techniques to predict changes in recruiter behavior in response to a change in mission. The results indicate that the effect of an increase in mission depends on the probability of making mission. More specifically, in a market where recruiters have a high probability of making mission, increasing the mission increases recruiter effort and expected enlistments. However, in a market where recruiters have a low probability of making mission, increasing the mission reduces recruiter effort. This result suggests that missions per recruiter should be based on supply of potential enlistments, with missions per recruiter higher in better markets.

The results also show how recruiter behavior will change in response to changes in the level of uncertainty about the number of recruits signed, given the level of effort. Specifically, increasing the level of uncertainty increases recruiter effort in circumstances with either a very high or a very low probability of making mission. In the intermediate range, increasing the level of uncertainty reduces recruiter effort. This suggests that in markets where recruiters have a fairly good chance of making mission, the reward structure should be based on team performance.[2]

## Simulations

We illustrate the comparative-static predictions described above by simulating recruiter effort under varying missions and varying levels of uncertainty. Table B.1 shows the parameter values used in the simulations, which were chosen to illustrate the range of theoretical possibilities. Thus, the numerical results should be interpreted in terms of qualitative patterns rather than as quantitative predictions. (Quantitative predictions based on analysis of data for FYs 1999–2001 are reported in Table 5.3.) Table B.2 shows the results from the simulations under the three uncertainty scenarios for a wide range of missions per recruiter. The results illustrate the comparative-static predictions just discussed. First, in all three scenarios, expected enlistments initially increase as missions are increased; however, if missions are set too high, expected enlistments decline as missions increase. Second, increasing the level of uncertainty increases expected enlistments if missions are either very low or very high. More specifically, expected enlistments are the highest in the "high uncertainty" case for missions that are either below 4 or above 14. Third, the optimal mission per recruiter depends on the level of uncertainty. For example, expected enlistments are maximized at a mission of 13 in the "high uncertainty" case and at a mission of 14 in the "low uncertainty" and "medium uncertainty" cases. Finally, the results show that if missions are set at their optimal levels, then reducing the level of uncertainty increases expected enlistments. The policy implications of these results are discussed in the body of the report.

---

[2]  We emphasize that the analysis here implicitly assumes that recruiters in a team can monitor each other so that the incentives to free ride on the effort of other team members are minimal, or that the increase in efficiency from a team-based organization offsets any losses in production due to free-riding.

Table B.1
Recruiter Responses to Uncertainty, Model Parameters

| Parameters | Scenarios, Level of Uncertainty | | |
|---|---|---|---|
| | High | Medium | Low |
| Reward for making mission | 400 | 400 | 400 |
| Cost of effort | $C(e) = e^2$ | $C(e) = e^2$ | $C(e) = e^2$ |
| Production function | $y = e + \varepsilon,\ \varepsilon \sim N(0, \sigma)$ | $y = e + \varepsilon,\ \varepsilon \sim N(0, \sigma)$ | $y = e + \varepsilon,\ \varepsilon \sim N(0, \sigma)$ |
| Sigma ($\sigma$) | 6 | 5 | 4 |
| Range of missions | 1–18 | 1–18 | 1–18 |

Table B.2
Simulation Results: Recruiter Responses to Changes in Missions and Uncertainty

| Missions per Recruiter | High Uncertainty | Medium Uncertainty | Low Uncertainty |
|---|---|---|---|
| 1 | 7.455 | 7.270 | 6.849 |
| 2 | 8.028 | 7.919 | 7.569 |
| 3 | 8.601 | 8.573 | 8.298 |
| 4 | 9.172 | 9.231 | 9.034 |
| 5 | 9.737 | 9.891 | 9.777 |
| 6 | 10.294 | 10.549 | 10.524 |
| 7 | 10.838 | 11.205 | 11.273 |
| 8 | 11.364 | 11.855 | 12.024 |
| 9 | 11.865 | 12.496 | 12.776 |
| 10 | 12.331 | 13.126 | 13.526 |
| 11 | 12.746 | 13.737 | 14.273 |
| 12 | 13.083 | 14.324 | 15.015 |
| 13 | 13.283 | 14.875 | 15.750 |
| 14 | 13.172 | 15.370 | 16.474 |
| 15 | 0.812 | 0.200 | 0.018 |
| 16 | 0.466 | 0.102 | 0.007 |
| 17 | 0.273 | 0.051 | 0.002 |

NOTE: Expected enlistments in a three-recruiter station.

# References

Chowdhury, Jhinuk, "The Motivational Impact of Sales Quotas on Effort," *Journal of Marketing Research*, Vol. 30, February 1993, pp. 28–41.

Darmon, Rene Y., "Selecting Appropriate Sales Quota Plan Structures and Quota-Setting Procedures," *Journal of Personal Selling & Sales Management*, Vol. 17, No. 1, Winter 1997, pp. 1–16.

Dertouzos, James N., and Steven Garber, *Human Resource Management and Army Recruiting: Analysis of Policy Options*, Santa Monica, Calif.: RAND Corporation, MG-433-A, 2006. As of August 17, 2007:
http://www.rand.org/pubs/monographs/MG433/

Oken, Carole, and Beth J. Asch, *Encouraging Recruiter Achievement, A Recent History of Military Recruiter Incentive Programs*, Santa Monica, Calif.: RAND Corporation, MR-845-OSD/A, 1997. As of August 17, 2007:
http://www.rand.org/pubs/monograph_reports/MR845/

Tubbs, Mark E., "Goal Setting: A Meta-Analytic Examination of the Empirical Evidence," *Journal of Applied Psychology*, Vol. 71, No. 3, 1986, pp. 474–483.

U.S. Census Web site. As of August 23, 2007:
https://askcensus.gov